# Principles of Tropical Horticulture

# Principles of Tropical Horticulture

Editor

**Babasaheb Desai**

scitus
academics

**Principles of Tropical Horticulture**

Edited by **Babasaheb Desai**

ISBN: 978-1-68117-090-9

Library of Congress Control Number: 2015935489


SCITUS Academics LLC,
www.scitusacademics.com
Box No. 4766, 616 Corporate Way,
Suite 2, Valley Cottage,
NY 10989



Printed in United States of America on Acid Free Paper ∞

# Contents

# Preface

Tropical horticulture is a branch of horticulture that studies and cultivates plants in the tropics, i.e., the equatorial regions of the world. "TropHort" is an abbreviation for Tropical Horticulture. Tropical Horticulture covers plants such as perennial woody plants (arboriculture), ornamentals (floriculture), vegetables (olericulture), and fruits (pomology) including grapes (viticulture). The origin of many of these crops is not in the tropics but in temperate zones. Their adoption to tropical climatic conditions is an objective of breeding. Many important crops, however, are indigenous to the tropics. The latter embrace perennial crops such as oil palm, vegetables including okra, field crops such as rice and sugarcane, and particularly fruits including pineapple, banana, papaya, and mango. A plant will interact with other plants, often to different species and other organisms to create a balanced community. Ecology is the study of this balance. In growing plants for our own ends we have created a new type of community which creates problems – problems of competition for the environmental factors between one plant and another of the same species, between the crop plant and a weed, or between the plant and a pest or disease organism. These latter two competitive aspects create the need for crop protection.

**Editor**

Chapter 1

# Vegetation Composition of Old Extensive Green Roofs (From 1980s Germany)

Christine E Thuring and Nigel Dunnett

Department of Landscape, Floor 9, The Arts Tower, Western Bank, Sheffield, South Yorkshire S10 2TN, UK

# ABSTRACT

## Introduction

Since their development in the late 1970s in Germany, extensive green roofs (EGR) have become increasingly popular as mitigation tools for urban environmental issues around the world. EGRs are planted with select species, which ensure consistent cover and performance over time. This research presented herein is part of a systematic re-evaluation of EGR technology since the German industry began.

## Methods

Given the opportunity to access a small sample of old EGRs installed over 20 years ago in south-west Germany, this research surveyed the vegetation and substrate with an interest in describing these parameters with time-through-space substitution.

## Results

Similar to previous studies, this preliminary work found correlations between roof age with vegetation (cover abundance and species diversity) and substrate properties (e.g., depth, organic content, pH, and nutrients). Roof age had positive relationship with soil organic content ($C_{org}$), and negative relationships with substrate depth and soil pH. These soil variables are inter-related, as shallow acidic substrates create unfavourable conditions for decomposition and thereby the accumulation of duff. Substrate variables correlated with EGR vegetation, suggesting a trend of simplified species composition over time. Indeed, $C_{org}$ had a negative relationship with cover and species diversity of most life forms; only Sedum species had positive associations with $C_{org}$.

## Conclusions

Considering the dynamics associated with shallow mineral substrates, and the greater floristic diversity of younger roofs, simple Sedum-based vegetation may represent a steady state for conventional EGRs.

# INTRODUCTION

Growing plants on roofs is an ancient concept common to many cultures and climates. Hanging (or roof) gardens were used by the ancient Greeks as personal sanctuaries to honour the god Adonis, while the Aztecs used them for urban agriculture and amenity (Arhendt, 2007). Grass (or sod) roofs have a long and global lineage, too, often for areas lacking building materials yet requiring insulation from exposure and extreme climates (e.g., Scandinavia, sub-Saharan

Africa, American Midwest) (Adler, 2005; Arhendt, 2007; Grant, 2006). Grass roofs are often colonised by spontaneous vegetation that closely represents the flora of the region. In the late 18th to mid-19th centuries, tar-paper-gravel (TPG) roofs became popular in many German cities (e.g., Berlin, Göttingen, Osnabrück, Karlsruhe) to stop the spread of fire and provide insulation (Köhler and Poll,2010). TPG roofs were standardised per city: Göttingen TPGs had 50 mm (each) of sand and gravel (Bornkamm, 1961), while Berlin TPGs comprised a thin layer of sand topped by 100–150 mm of gravel with loam (Darius and Drepper, 1983). The vegetation that colonised these early systems and/or which developed following sowings in the 1980s, range from xeric *Sedo-Scleranthetea* communities on shallow depths to grassy *Festuco-Brometea* communities on deeper substrates (Bornkamm, 1961; Bossler and Suszka, 1988; Buttschardt, 2001; Poll, 2008).

Today's modern extensive green roofs (EGRs) differ from these predecessors in that they are thin, lightweight systems designed as socio-environmental solutions for the urban environment (Krupka,1992). Developed in Germany, EGRs emerged in the 1970s as part of that era's greater Green Movement, which also included materialization of the Green Party, Greenpeace, etc. (Galtung,1986). In the spirit of the times, vegetated roofs were also seen as opportunities to reconnect urban dwellers with nature (Minke and Witter, 1983). Due to constraints imposed by roof loading and low maintenance requirements, EGRs stand apart from earlier green roof types by their system build-ups (Figure 1) which balance environmental function and performance with a standardised and economically sustainable green roof market.

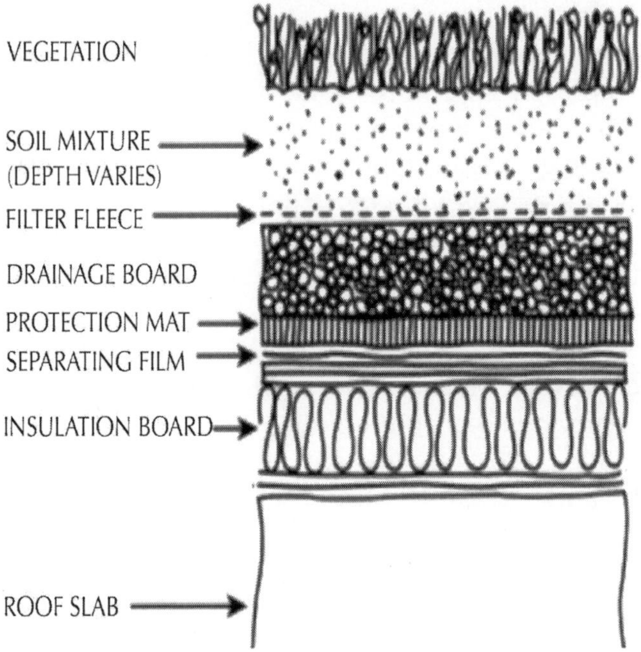


**Figure 1:** Example of extensive green roof system build-up. Image courtesy of Grant et al. (2003).

In the meantime, Earth's human population continues to grow (UNFPA, 2011) while becoming increasingly urbanised (UNFPA, 2007). Along with other forms of green infrastructure, green roofs are recognised as valuable technologies for alleviating the environmental impacts of urbanisation and restoring ecosystem services, which include potable water, clean air, crop pollination, and social well-being, among others (CBD, 2012; Millennium Ecosystem Assessment, 2005a). Because of their capacity to support biodiversity, which is the keystone of ecosystem services (Millennium Ecosystem Assessment, 2005b), green roofs can contribute to a healthy, resilient, and equitable future. Concurrently, green roof markets continue to grow in Germany (FBB, 2012) and in other parts of the world [e.g., Peck, 2012.

Modern EGRs comprise at least two layers – vegetation and growing substrate – but usually multiple layers are used (e.g., root barrier, protection mat, drainage layer). By definition, extensive systems are shallower than 200 mm while intensive roofs (or roof gardens) are deeper (FLL, 2008). Table 1 summarises the types of vegetation for the range of EGR depths. Being intended for amenity, intensive green roofs have higher loading capacities than EGRs, and their deeper depths permit larger plants and more species, not to mention park-like features such as ponds and benches (Osmundson, 1999; Weiler and Scholz-Barth, 2009). Due to issues associated with poor drainage, single-layer systems should be limited to roofs with a minimum 2% slope (Krupka, 2006). On multiple-layer systems, the drainage layer is designed to move excess water towards roof drains in order to minimise water logging and hydrostatic load (Kolb and Schwarz, 1999). Theoretically, drainage layers with storage cups can provide moisture to plants during periods of drought but, unless the cups are filled with granular infill, the substrate is effectively separated from the stored water by an air gap, which prevents capillary action. Therefore, the only way by which plants can access such moisture is through evaporation from the stored water into the substrate (Vesuviano, 2013).

**Table 1:** Different depths support different vegetation forms

| Depth (mm) | Vegetation form |
|---|---|
| 40–100 | Moss, sedum |
| 50–110 | Sedum, moss, herbaceous |
| 100–180 | Sedum, herbaceous |
| 150–250 | Grass, herbaceous |

Adapted from FLL (2008, p. 43).

Thuring and Dunnett

Thuring and Dunnett *Ecological Processes* 2014 3:4, doi:10.1186/2192-1709-3-4

The first guidelines for the planning, construction, and maintenance of green roof systems were published in 1982 by the German Landscape Development and Landscaping Research Society, or the FLL (Forschungsgesellschaft Landschaftsentwicklung Landschaftsbau, 2008). These technical standards have been essential to the development and sustained growth of the German green roof industry and market (Ansel et al., 2011).

For EGR plant selection, early green roof pioneers took ecological reference from climax and sub-climax plant communities that occur in environmental conditions analogous to those of roofs (e.g., montane, dry grasslands). Hardy and drought-tolerant taxa, including native and non-native species as well as ornamental cultivars that are suited to extreme exposures were trialled and tested, and now serve as the foundation of many EGR species lists (Kolb et al., 1983; Kolb and Schwarz, 1999). One exceptionally well-suited plant community, the *Sedo-Scleranthetea*, is defined by poor grassland consisting primarily of low-growing herbs, short-culmed and thin-leaved grasses, mosses and lichens (Ellenberg, 1986). The herbs are represented predominantly by succulents, in particular Sedum species and winter annuals, which explains why Sedum species have become synonymous with EGRs.

EGR substrates are 70% to 90% mineral (by volume) (Kolb and Schwarz, 1999), and engineered 'soil-less' media are commonly used for rooftop applications for a variety of reasons. For one, the high porosity of mineral substrate promotes excellent drainage and oxygenation for the root environment, but also holds sufficient water to support plant growth between rain events (Handreck and Black, 2010). Furthermore, mineral aggregates resist compression and shrinkage, thereby maintaining medium structure and further promoting drainage and aeration (Dunnett and Kingsbury, 2004). It is important to recall that these standards were developed in the continental climate of central Europe, and that EGRs in other climates often require different criteria [e.g., Williams et al., 2010.

The amount of rainfall retained by green roofs is of particular interest to cities with aging stormwater infrastructure because of their function of retaining and detaining runoff. Depth strongly influences substrate stormwater performance, as deeper substrates have greater storage capacity (Mentens et al., 2006; VanWoert et al., 2005). In spite of their shallow substrates (<150 mm), EGRs have measured cumulative annual retentions of 70% in northern Germany (Liesecke, 1995), 60% in North Carolina (Moran et al., 2004), and 50% in northern England (Stovin et al., 2012). In addition to depth, retention values depend on the intensity and duration of a rain event, local climate, vegetation, and roof slope (DeNardo et al., 2005; Dunnett et al., 2008b; VanWoert et al., 2005; Villarreal and Bengtsson, 2005; Yio et al., 2013). Most hydrologic models are based on relatively young green roofs or test facilities and, other than one company's proprietary research station (Uhl et al., 2003), the evaluation of green roof hydrological performance over time is largely unknown.

In addition to their benefits of mitigating stormwater runoff, EGRs are promoted in many cities around for the world for improving air quality (Speak et al., 2012), reducing the urban heat island (Dimoudi and Nikolopoulou, 2003; Takebayashi and Moriyama, 2007; Wong et al., 2003), improving thermal insulation (Simmons et al., 2008), mitigating low frequency noise (Connelly and Hodgson, 2013), supplementing urban green infrastructure (Grant, 2012), enhancing urban biodiversity (Brenneisen, 2006; Gedge and Kadas, 2004; Grant et al., 2003) and much more (Francis and Lorimer, 2011; Oberndorfer et al., 2007).

Ecological designs, particularly through manipulations of the substrate, can transform EGRs from technical systems into heterogeneous ecosystems. Deeper depths offer more diverse vegetation that can in turn provide valuable foraging resources for urban bees (Tonietto et al., 2011). Diversifying substrate materials and creating mounds can offer refuge to invertebrates from extreme temperatures (Buttschardt, 2001). Such variations of microhabitat correlate with invertebrate species richness (e.g., spiders, beetles) (Brenneisen, 2003; Kadas, 2011; Mann, 1996). In Basel, Brenneisen (2009) showed that structurally diverse substrates and a stable mosaic of gaps in the vegetation will allow rare, even extirpated, invertebrates to colonise shallow EGRs, although this has yet to be tested and achieved in other places. The provision of habitat resources, like perches and food and water supply, can accommodate the breeding needs of ground-nesting birds (Baumann, 2006; Fernandez-Canero and Gonzalez-Redondo, 2010).

A small body of plant community ecology studies has applied phytosociological methods to green roofs in the attempt of classifying the vegetation into plant communities. The vegetation of old TPG roofs approximates the composition of some natural plant communities, and supports a novel plant community, the *Poetum anceptis-compressae* (Typical Poa meadow) (Bornkamm, 1961; Bossler and Suszka, 1987; Buttschardt, 2001; Darius and Drepper, 1983; Poll, 2008). Although it approximates the *Sedo-Scleranthetea*, EGR vegetation does not classify satisfactorily into plant communities mainly because of introduced species and deficient seed rain (Buttschardt, 2001; Poll, 2008).

Other than the ecological research mentioned above, most green roof research is engineering- or horticulture-oriented with the goal of understanding performance and function under different conditions (Blank et al., 2013). The requirements of experimental controls and replication mean that green roof research is usually conducted on mock-ups or platforms (Dvorak and Volder,2010). If actual green roofs are studied, they are rarely older than 10 years and the timeframe for most green roof successional research is relatively short [e.g., Bates et al., 2013; Rowe et al.,2012. However rare, long-term observations of ecological phenomena and the consistent and reliable accumulation of long-term synoptic datasets are essential to understanding how natural systems work (Callahan, 1984; Franklin et al., 1990; Likens, 1989). In any case, the study of EGRs as ecosystems subject to the laws and processes of nature, as defined by ecological theories and principles, has hardly been touched upon (Cook-Patton and Bauerle, 2012).

If EGRs can serve as green infrastructure solutions for an increasingly urbanising planet (Grant, 2012; Millennium Ecosystem Assessment, 2005a), then understanding their long-term performance is crucial. Furthermore, understanding the factors which affect species richness on EGRs over time can help towards more informed designs and specifications for biodiversity in the urban context. This research surveyed the vegetation and substrate on some of the oldest EGRs in south-west Germany and bolsters its limited sample size with the Buttschardt (2001) dataset, which had been analysed differently. Based on the species poverty of TPG roofs, we hypothesised that floristic diversity on EGRs also becomes more homogenous over time. With reference to competition theory and other green roof research, we hypothesised that very shallow substrate depths (≤80 mm) would provide higher species diversity compared with deeper depths (≥100 mm).

# METHODS

The development of vegetation on EGRs has been conjecture at best due to issues such as accessibility and deficient scientific method. An EU-funded Industry-Academia Partnership (Marie Curie IAPP) helped to close this gap (Green Roof Systems Project, 2013). One of the earliest EGR system manufacturers, the industry partner (ZinCo GmbH) helped gain access to six of its oldest EGRs (Roofs 1–6). Three non-ZinCo roofs were included, as well. The sampling region was confined to the Stuttgart region (south-west Germany), which is typified by a continental climate.

Nine EGRs were surveyed over the course of two growing seasons (2010 and 2011) (Table 2). Several roofs feature prototypic EGR systems and/or materials, some of which have become commonplace to the green roof industry while others have not been taken up. Two of the pitched roofs (Roofs 2 and 4), for example, were custom built because such greening had not been done before, while the pre-grown Styrofoam modules on Roofs 5 and 6 did not last long on the German market. In spite of the differences in location, age, area and slope, the roofs are all based by typical EGR substrates (e.g., mineral: organic ratio), and field observations confirmed the use of multiple-layered systems.

**Table 2:** Summary of the nine extensive green roofs in order of age at time surveyed

| Roof # | Roof name | Age at survey | Year installed | Area ($m^2$) | Slope (°) | # quadrats | # species |
|---|---|---|---|---|---|---|---|
| 1 | FH Nürtingen | 23 | 1987 | 258 | 0 | 12 | 32 |
| 2 | Römermuseum, Köngen | 23 | 1987 | 350 | 17 | 18 | 9 |
| 3 | Pliensaufriedhof, Esslingen | 33 | 1977 | 500 | 0 | 15 | 11 |
| 4 | Gärtnereihof Tübingen | 24 | 1986 | 2,160 | 15 | 16 | 21 |
| 5 | Verkehrsbetrieb Area 1 | 25 | 1986 | 1,860 | 0 | 14 | 21 |
| 6 | Verkehrsbetrieb Area 2 | 25 | 1986 | 2,064 | 0 | 14 | 21 |
| 7 | Stuttgart Rathausgaragedach, PV | 21 | 1990 | 1,300 | 0 | 15 | 26 |
| 8 | Stuttgart Rathausgaragedach, low | 21 | 1990 | 1,000 | 0 | 14 | 30 |
| 9 | Killesberg Dach | 20 | 1991 | 450 | 30 | 18 | 23 |

Thuring and Dunnett

Thuring and Dunnett *Ecological Processes* 2014 3:4, doi:10.1186/2192-1709-3-4

Although initial data was not available, the roofs surveyed were all built in the early years of the German green roof industry and would have adhered to the early FLL standards. Verbal information suggests that all roofs were regularly maintained in their early years, which would have included weeding and possibly mowing and fertilising. According to personal correspondence with roof contacts, maintenance contracts were not renewed in the early 1990s, which means that the roofs had not been maintained in about 10 years at the time of survey.

Little written baseline information was available for the roofs surveyed. Since German companies/institutions typically purge any documentation over 10 years old, only sites with staff members interested in the green roofs had retained any related documents (Roofs 1, 4, 9). Some roofs had technical drawings that included reference to species lists and substrate depths (Roofs 2, 7, 8). Unfortunately, original substrate depth, details of substrate composition and species lists cannot be known for certain. That being said, since the roofs were built in the time of the FLL guidelines, they likely adhered to those specifications, meaning that soil pH was between 6.5 and 8.0, soil organic content below 65 g/L, and substrate depth less than 200 mm (FLL, 2008). These results mainly present a descriptive snapshot of what is present today.

## Data Collection

The traditional tool of quantitative plant ecology, a 1-$m^2$ quadrat (Braun-Blanquet, 1932; Mueller-Dombois and Ellenberg, 1974), was used for floristic description, specifically for cover abundance records. With the goal of describing EGR vegetation, primary survey methods with ecological objectives were deemed appropriate for including the diversity of roof constructions since the intent was to correlate local variation in vegetation composition with variation in environmental factors (van der Maarel, 2005). Floristic sampling occurred from early June to mid-July and substrate and biomass harvest in autumn. Each roof was sampled once and is represented by between 12 and 18 quadrats; this number depended on the type of roof, its surface area, and the vegetation.

Quadrat placement was determined by site conditions, vegetation homogeneity, environmental gradients, and the statistical requirements of sampling (Kent and Coker, 1994). A small roof with uniform vegetation (Roof 1) was sampled by systematically locating points at regular intervals, for example, while roofs with heterogeneous vegetation were sampled using stratified methods that clustered major sources of variation. In those cases, stratified methods first defined physiognomic groups (e.g., Sedum cover versus shrubby mounds) and then sampled the predominant, homogeneous vegetation (Roofs 3, 5–8).

Similarly, pitched roofs (Roofs 2, 4, 9) were sampled using systematic transects, which permit the description of maximum variation across an environmental gradient over the shortest distance (Kent and Coker, 1994). The diversity of roof constructions from this primary survey was united for analysis because of the limited sample size but also through reference to the methodological approach of Buttschardt (2001), whose surveys included two pitched roofs in a sample of less than ten EGRs.

Above-ground cover abundance by plant species was measured at percent cover (%) per $m^2$ for individual species and also by amalgamated physiognomic life form groups (Table 3). Although they qualify as forbs, succulents (e.g., Sedum species) were separated into their own group because of their unique structures and strategies. for the full species list and species frequency on the nine roofs. Certainly, grouping species has the drawback of masking the variation of ecological strategy within each group, but it can also reveal important environmental factors influencing the structure of the vegetation. Likewise, treating large numbers of species individually can hinder the analysis and interpretation of the broader functional aspects of the vegetation (van der Maarel, 2005).

**Table 3:** Species were grouped into physiognomic life form groups

| **Life form** | **Vegetation type** |
|---|---|
| Woody | Trees, shrubs, sub-shrubs |
| Grasses | Grasses |
| Succulents | Succulent and crassulacean species |
| Forb | Herbaceous flowering plants |
| Bulb | Bulbous flowering plants |
| Cryptogam | Mosses and liverworts |

Thuring and Dunnett
Thuring and Dunnett *Ecological Processes* 2014 3:4, doi:10.1186/2192-1709-3-4

Each sampling plot received a record for substrate depth, taken as the mean of three measurements per quadrat. Depth measurements typically ranged from the substrate surface to the filter sheet separating substrate from the drainage layer below. Taken together, each roof therefore received a mean value for substrate depth. Interestingly, in spite of the different forces one would expect, substrate depth on pitched versus flat roofs was similar for both the mean and the measured range across each roof (data not shown); of course, the absence of original data prevents any meaningful conclusions. Overall, we would expect pitched roofs to have deeper depths at the base of the slope and shallower depths at the top, supporting an accordingly stratified vegetation (Table 1).

Finally, excepting the Killesberg site, eight roofs were sampled for physical and chemical substrate properties using a (10 cm) soil corer (Firma Schwab, Waidhofen). One or two cores were taken per quadrat, all vegetation removed, and core profiles photographed before being united into a single, 20 L sample (15 L are required for analysis, plus 5 L retain sample). Cored gaps were re-filled with a commercially available green roof substrate, 'Steinrosenflur'. The University of Hohenheim LA-Chemistry Laboratory conducted soil analyses in adherence with the FLL guidelines.

## Data Analysis

The information used for analysis here includes cover abundance (%) of life form groups as well as species diversity; roof age (age at time of survey); soil depth (mm); soil nutrients (nitrogen, phosphorus, potassium and magnesium) (as measured in mg/L); soil organic content (g/L); and soil pH (as measured by $CaCl_2$). As a first step, the relationships between these multiple variables are investigated using Spearman's Rank Order Correlation (*rho*), which can describe the strength and direction of relationships. This non-parametric solution was chosen because some of the data was ordinal and most of it was not normally distributed. Preliminary tests were performed to ensure no violation of the assumptions of normality, linearity and homoscedasticity.

# RESULTS AND DISCUSSION

Given the small sample size and the lack or incomplete quality of original documentation, we can only speculate about how the vegetation and substrate properties develop on EGRs over time. In the results that follow, it is important to note that the context of causation and interlinkage amongst variables is inadequately known; there may be spurious correlations or significant relationships that are influenced more by lurking variables rather than the two variables identified. To address the possibility of spurious correlations, the results and discussion may serve for consideration of the scenarios suggested but shall not be treated as conclusions. Since soil properties fun

## Roof Age and Soil Properties

The older EGRs had lower soil pH and less substrate depth, less biomass, and greater soil organic content than younger roofs. Some of these effects are clearly related and do not require elaboration (e.g., less soil depth = less biomass), while others are of particular relevance to long-term green roof performance.

## Decreased Soil PH Over Time

Although not significant, the roofs surveyed in this study concur with findings from others in suggesting that soil pH declines over time. Of the eight roofs sampled for soil, pH values ranged from 5.2 to 7.2. Only the two youngest roofs fell within the recommended range of 6.5 to 8.0 [according to FLL (2008)]. A similar study in Karlsruhe, of EGRs between 3–8 years old at time of survey (installed between 1992–1997), recorded soil pH values between 5.8 and 7.6, of which two (7 years old) fell below the FLL standard (Buttschardt, 2001). Of 10 EGRs sampled in Hannover, Schrader and Boening (2006) also found that pH values for the older roofs all fell below the FLL recommendation. In any case, soil pH was lower on the older EGRs (8–12 years old at time of survey; installed 1990–1994) compared to the younger roofs (3–4 years old; installed 1998–1999). Köhler and Poll (2010) observed differently in Berlin because EGRs there were systematically amended with lime as a response to acid rain in the 1980s.

In fact, a number of controlled studies have noted significant declines in soil pH of EGR substrates over time. In a systematic test of 23 mineral substrates (no organic matter), Jauch and Fischer (2000) found that all trended towards acidification over the 7-year study period. Lava substrates fell from between pH 7.5 and 8.2 (in 1993) to 5.7 and 6.2 (by 2000); expanded clays from between pH 6.7 and 9.6 (1993) to 5.1 and 6.0 (2000); and expanded shale from between pH 8.1 and 9.0 (1993) to 4.8 and 6.3 (2000). The most stable substrates tested were those composed of brick, with pH values declining from between 7.0 and 9.0 (1993) to 5.8 and 7.7 (2000). Another study of twelve substrates over a period of up to 16 years, which included the same aggregates described above plus others, noted that pH drops significantly in the first years after installation (Liesecke, 2006). The apparent drop in pH suggests that the materials comprising EGR substrates lack sufficient buffering capacity. These effects are discussed in detail further on.

## Less Depth over Time

A preliminary correlation analysis suggests that substrate on older EGRs was significantly shallower than on the younger roofs surveyed ($rho = -0.481$, $P < 0.000$, $n = 134$). Since EGRs are relatively shallow to begin with, and given the lack of long-term performance research on these systems, such a phenomenon would have great consequences for emerging markets that employ minimal FLL guidelines because regionally tested standards have not yet been developed (Snodgrass and McIntyre, 2010). Recalling the caution of spurious relationships as expressed above, and the limitations of sample size in this study, this effect nevertheless deserves careful consideration.

Table 4 shows the surveyed EGRs arranged by ordinal age groupings of mean depth (mm) per roof. These values, illustrated in Figure 2, suggest a linear decrease in mean substrate depth over time. If the data point from the oldest roof surveyed (Pliensaufriedhof) is considered an outlier, a negative trend becomes clearer. That roof supports dense Sedum cover that is typical EGR vegetation, but it was originally installed as a semi-intensive system (in 1978) and would have received more substrate than its younger counterparts.

**Table 4:** Mean substrate depth on nine extensive green roofs in order of increasing roof age

| Roof age grouping | Roof # | Roof name (in order of age at time of survey) | Mean substrate depth (mm) |
|---|---|---|---|
| 20–21 | 7 | S-Rathausgaragedach, PV | 75.4 |
| | 8 | S-Rathausgaragedach, low | 69.8 |
| | 9 | Killesberg Dach | 84.7 |
| 22–23 | 1 | FH Nürtingen | 72.3 |
| | 2 | Römermuseum, Köngen | 76.6 |
| 24–25 | 4 | Gärtnereihof Tübingen | 61.5 |
| | 5 | Verkehrsbetrieb Area 1 | 52.8 |
| | 6 | Verkehrsbetrieb Area 2 | 58.1 |
| >26 yr | 3 | Pliensaufriedhof, Esslingen | 61.6 |

Thuring and Dunnett

Thuring and Dunnett *Ecological Processes* 2014 3:4, doi:10.1186/2192-1709-3-4

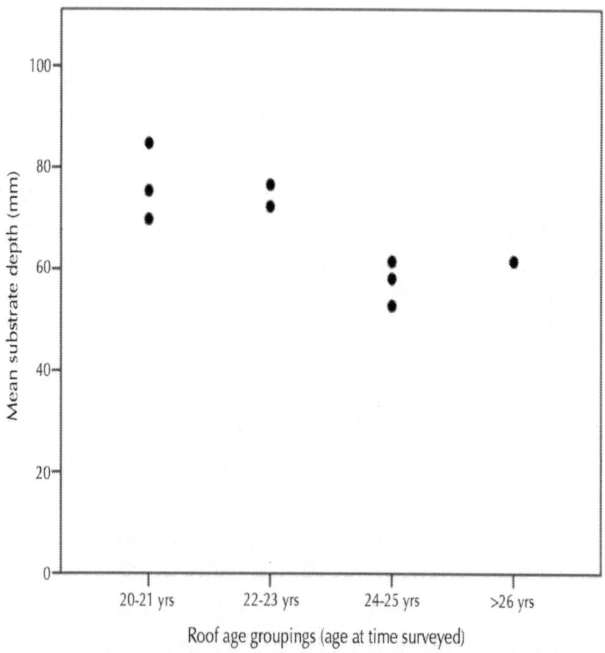


**Figure 2:** Mean substrate depth decreases on extensive green roofs over time.

A Kruskall-Wallis test revealed a statistically significant difference in substrate depth across the four age groupings (depth 1, n=46: 20–21 years; depth 2, n=29: 22–23 years; depth 3, n=44: 24–25 years; depth 4, n=15: >26 years), $\chi^2$ (3, n=134)=42.74, $P<0.01$). The two oldest roof age categories had small means (39.7 mm, 53.3 mm, respectively), especially by contrast with the younger age groups, which had mean depths greater than 86 mm. In spite of the small sample size, and the variability amongst roofs, these results strongly suggest that EGR substrate depth decreases over time, although it is unclear how or when this would occur.

Both the FLL standards and industry practice in Germany are designed to rule out compaction from point of installation. First, the DIN 18127 laboratory standard (Proctor Test) for substrate manufacture ensures that a substrate has already factored compression into its ordered volume. Further, in the case of EGR substrates delivered by silo trucks (which compromise particle size distribution due to shattering) (Roth-Kleyer, 2006), the FLL (2008) recommends maintaining the prescribed granulometric distributions by including greater proportions of large particles into those blends. In addition to the FLL, EGR system and/or substrate providers will often calculate substrate volume for installation using 'settlement factors' of between 1.1 and 1.25, depending on the substrate (e.g., ZinCo GmbH, 2013, p. 15).

Although most of these roofs were installed in the early days of the EGR industry, they may still be considered representative of conventional German EGRs. The oldest roof sampled (Roof 3, Pliensaufriedhof) preceded the first FLL guidelines by four years, but it features a system build-up that remains commonplace to the industry so would presumably embody certain standards. The roofs surveyed would not have been installed using silo trucks, either, which only became prevalent after the industry had developed sufficiently; by 2002, 70% of EGR installations in Germany used blower trucks (Roth-Kleyer, 2002).

Substrate depth fundamentally affects EGR vegetation, whether in terms of establishment and growth (Durhman et al., 2007; Getter and Rowe, 2009; Rowe et al., 2012; Thuring et al., 2010), species dominance and cover diversity (Dunnett et al., 2008b; Emilsson and Rolf, 2005; Nagase and Dunnett, 2010), or survival across challenging seasons (Boivin et al., 2001; Getter and Rowe,2007). Shallower depths also inhibit the growth of taller vegetation (Table 1) which together compromise hydrological performance (Dunnett et al., 2008a) and other benefits like the cooling of buildings in hot seasons (Sailor, 2008; Santamouris et al., 2007).

## More Soil Organic Content over Time

The older roofs surveyed typically had more soil organic content ($C_{org}$) than younger roofs. With reference to the FLL specification that $C_{org}$ should be no higher than 65 g/L, four of the eight roofs were in excess (FH Nürtingen: 72 g/L; Pliensaufriedhof: 189 g/L; Köngen: 126 g/L; Tübingen: 79 g/L), two were close to the limit (Verkehrsbetrieb A1: 60 g/L; Rathausgarage lower: 61 g/L) and two were acceptable (Verkehrsbetrieb A2: 25 g/L; Rathausgarage PV: 49 g/L). As mentioned above, the roofs sampled probably adhered to the FLL guidelines and the same would apply to substrate composition. Vegetation surveys of EGRs in Karlsruhe recorded 'high to very high' $C_{org}$ of which measured values did not differ significantly from those of old, spontaneously vegetated TPG roofs (Buttschardt, 2001). Köhler and Poll (2010) in Berlin similarly observed that $C_{org}$ on 20-year-old EGRs was not significantly different from 100-year-old TPGs.

Organic content in EGR substrates is specified to be low because organics retain moisture better than mineral substances (influencing roof loads), while oxidation, compression and other forms of weathering limit their longevity (Kolb et al., 1983). Similar to 'soil-less' potting media, mineral-rich EGR substrates are designed to provide comparable moisture holding properties to organic-rich mixes through specific particle size distributions (FLL, 2008; Miller, 2003). The limited nutrient content of high-mineral EGR substrates gives the desired green roof vegetation a competitive edge over ruderal volunteer species, which typically require more fertile soils (Kolb and Schwarz, 1999). Nutrient-holding capacity is an important consideration for organic material on EGRs, too, particularly with respect to runoff quality and surface water pollution (Emilsson and Rolf, 2005).

In their study of carbon sequestration on twelve Sedum-based EGRs (ranging from 1 to 6 years in age and 2.5 to 12.7 cm in depth), Getter et al. (2009) found that the three shallowest roofs (2.5 cm) increased in soil carbon with respect to age. Their parallel, plot-based study (6 cm) found that 100 g $C*m^{-2}$ was sequestered by the substrate over two growing seasons. Liesecke (2006) similarly observed, in twelve roofs sampled in northern Germany, that the four single-layer EGRs with expanded shale (5, 8, 11, 14 cm) had the highest $C_{org}$ and thickest duff layer after 12 years. Of those, the shallowest depths had the highest $C_{org}$ values. By contrast, the multiple-layered systems decreased in $C_{org}$ content. These results infer that decomposition by microbial activity is inhibited on shallow, mineral-rich EGRs, hence the accumulation of organic material in the form of thatch or duff. Liesecke (2006) also reported that the same four roofs had the highest values for water holding capacity and exceeded to nearly twice their specified loading weight. While beneficial for stormwater retention, this development can have serious repercussions if the roof structure is not designed for such loads. It is worth noting that single-layer EGR systems declined in popularity in Germany since that study.

A couple studies report that $C_{org}$ on EGRs sinks in the early years after installation and then increase after 10 years, and that soil formation, or pedogenesis, begins to occur after 20 years (Köhler and Poll, 2010; Schrader and Boening, 2006). Although one might expect a build-up of 'nutrient capital", the accumulation of organic matter in soil can actually lead to a decrease in nutrient availability over time (Hodgson, 1990). In semi-natural grasslands, the accumulation of dead plant material over the course of natural succession on nutrient-poor mineral substrates can lead to 'dramatic changes' in the soil (Berendse, 1998, p. 85). The increase in soil organic matter can set off a series of further developments in the soil, including increased nitrogen mineralisation (by a factor of 10 to 20 over as little as 50 years), increased soil moisture content, as well as a strong decline in soil pH and soil $Ca^{2+}$ content (*ibid*).

Decreasing soil pH, as suggested by this and other studies, can also lead to an accumulation of $C_{org}$ because of the effects of pH on microbial activity, which are essential to nutrient cycling and decomposition, as well as plant nutrient uptake (Berendse, 1998). Beneficial microorganisms have preferred pH ranges in which they function best: *Rhizobium* bacteria prefer >5; ectomycorrhizal fungi prefer 4 to 6 (some 7); endomycorrhizal fungi prefer 4.5 to 8; bacteria that convert ammonium to nitrate prefer >6; and bacteria that attack fungi prefer 6.5 to 7.5 (Handreck and Black, 2010). Roof exposure, substrate depth, and distance to nearest green space are the main limitations to colonisation of EGRs by soil fauna, which are limited to xero- and thermophilic pioneer species (Schrader and Boening, 2006).

Likewise, sufficient nutrient provision and strong plant growth combined with incomplete litter decomposition inevitably leads to humus accumulation. However, EGR substrates can only offer limited plant nutrition due to their high water permeability and shallow depths, not to mention low cation exchange capacity (Emilsson et al., 2007). Since mineralised nutrients are rapidly leached out of EGR systems (*ibid*), the application of slow-release fertiliser is part of the EGR maintenance protocol (FLL, 2008). Several studies propose that ecosystem services by green roofs can be enhanced by providing the conditions for diverse and persistent communities on EGRs – whether plant species and functional groups (Lundholm et al., 2010) or soil biota, like invertebrates (Kadas,2011) and fungi (McGuire et al., 2013).

## Roof Age and Vegetation Development

Some life forms are more interesting than others to this study. Grasses and forbs, for example, are of interest because they represent great potential for floristic diversity and thereby more complex opportunities for food webs and the habitat needs of animals and insects (Baumann, 2006; Kadas, 2011; Mann, 1996). Succulents are interesting because they are popular on species lists around the world (Snodgrass and Snodgrass, 2006). The most prevalent genus, Sedum, can typically tolerate extreme conditions yet perform well in favourable conditions, while also creating reliable roof vegetation cover, which absorbs moisture, prevents erosion, inhibits colonisation, and requires hardly any maintenance (Kolb and Schwarz, 1999). While we cannot know how plant cover has changed over time, succulents certainly had the most consistent cover of all the life forms surveyed here, whether as dominant species or as constant groundcover beneath taller herbs and grasses (data not shown).

## EGR Floristics Most Influenced By $C_{org}$ and Phosphorus

Cover and species diversity of all life forms were most strongly affected by substrate properties. $C_{org}$ had negative relationships with cover abundance and species diversity for all life forms except succulent cover which had a positive relationship. Soil phosphorus had similar effects as $C_{org}$, namely negative relationships with the exception of succulents' positive effects, while soil pH had negative relationships with succulents but positive relationships with other life forms.

Recalling the interactions between $C_{org}$, soil pH, and depth, Jauch and Fischer (2000) observed that spread and cover by Sedum vegetation was negatively influenced by the sharp decline in pH two years after installation. Liesecke (1998) also observed lower species diversity on older EGRs, reporting that these systems may become dominated by one or two succulents, a single herb, and one or two moss species. Soil pH influences plant nutrient uptake because essential mineral nutrients combine with other elements at the highest and lowest pH values and thereby limit availability for plants. Nutrient absorption is also dependent on pH since an acidic substrate (less than pH 7) inhibits cation exchange capacity (Handreck and Black, 2010). Emilsson et al. (2007) and several German studies have shown that high plant cover is difficult to achieve without continuously adding fertiliser.

## Next Steps

Further analyses will attempt to determine the role of site-specific environmental conditions to vegetation development, to explain how community-level dynamics like persistence versus colonisation proceed with time, and to characterise the physical and chemical properties of old EGR substrates. Increasing the sample size by amalgamating with other datasets may improve the power of this research; in the meantime, these results can serve as a point for discussion. More long-term monitoring of EGRs, with emphasis on vegetation and substrate development as well as hydrological function, will dramatically improve our understanding of these systems (Dunnett et al.,2008b; Rowe et al., 2012) and of urban ecosystems overall.

# CONCLUSIONS

While they may not permit causal explanations, a patent theme cannot be overlooked, namely that EGRs are dynamic ecological systems that respond to multivariate factors and are subject to change over time. This research suggests that shallow EGRs tend towards substrate acidification and accumulation of soil organic content, while also becoming shallower. These effects equally influence the composition of EGR vegetation, specifically in terms of life form cover and species diversity, whereby succulent cover represents a type of climax community. Floristic diversity on the roofs surveyed ranged from simple, low diversity Sedum roofs to consistent succulent cover beneath tall meadow vegetation. Although this study is limited by sample size and lacking baseline information, its preliminary observations are substantiated by similar studies.

# AUTHORS' CONTRIBUTIONS

CET jointly conceived and designed the study with ND, collected and analysed data and wrote the manuscript. ND supervised the project and gave conceptual advice. Both authors read and approved the final manuscript.

# ACKNOWLEDGEMENTS

Many thanks to colleagues at ZinCo GmbH, as well as other friends within the German green roof industry, and to IAPP Green Roof Team, without whose support this research would not be possible. Thanks to Dr. Oliver Pescott and Prof. Dr. Reinhard Böcker (University Hohenheim) for help in identifying mosses and graminoids. The author is indebted to harvest helpers (Trinity Pearce, Gianni Vesuviano, Claire Cannon, and Dr. Tobias Emilsson).

# REFERENCES

1. Adler P (2005) Die Dachbegrünung im Wandel der Zeit. Hochschule Nürtingen: Diplom (Master's).
2. Ansel W, Baumgarten H, Dickhaut W, Kruse E, Meier R (2011) Leitfaden Dachbegrünung für Kommunen – Nutzen – Fördermöglichkeiten – Praxisbeispiele. Nürtingen: Deutsche Bundesstiftung Umwelt.

3. Arhendt J (2007) Historische Gründächer: Ihr Entwicklungsgang bis zur Erfindung des Eisenbetons. Teil I. PhD (Dr.-Ing.). Berlin: Technische Universität Berlin.
4. Bates AJ, Sadler JP, Mackay R (2013) Vegetation development over four years on two green roofs in the UK. Urban For Urban Green 12:98-108
5. Baumann N (2006) Ground-nesting birds on green roofs in Switzerland: preliminary observations. Urban Habitats 4:37-50
6. Berendse F (1998) Effects of dominant plant species on soils during succession in nutrient-poor ecosystems. Biogeochemistry 42:73-88
7. Blank L, Vasl A, Levy S, Grant G, Kadas G, Dafni A, Blaustein L (2013) Directions in green roof research: a bibliometric study. Build Environ 66:23-28
8. Boivin MA, Lamy MP, Gosselin A, Dansereau B (2001) Effect of artificial substrate depth on freezing injury of six herbaceous perennials grown in a green roof system. Hort Technol 11:409-412
9. Bornkamm R (1961) Vegetation und vegetations-entwicklung auf Kiesdäechern. Vegetatio 10:1-24
10. Bossler S, Suszka B (1987) Vegetation und Substrat auf Dächern in Osnabrück. FH-Osnabrück: Diplomarbeit.
11. Bossler S, Suszka B (1988) Spontanvegetation auf Dächern in Osnabrück. Das Gartenamt 37:209-221
12. Braun-Blanquet J (1932) Plant sociology. In: Fuller GD, Conrad HS (eds) The study of plant communities/translated revised, New York/London: McGraw-Hill.
13. Brenneisen S (2003) The benefits of biodiversity from green roofs: key design consequences. Greening Rooftops for Sustainable Communities. Chicago, IL: Green Roofs for Healthy Cities.
14. Brenneisen S (2006) Space for urban wildlife: designing green roofs as habitats in Switzerland. Urban Habitats 4:27-36
15. Brenneisen S (2009) Ökologisches Ausgleichspotenzial von Extensiven Dachbegrünungen: Bedeutung des Ersatz-Ökotops für den Arten- und Naturschutz und die Stadtentwicklungsplanung, Liestal. Basel: Physiogeographie und Umweltwandel des Geographischen Instituts der Universität Basel.

16. Buttschardt TK (2001) Extensive Dachbegrünungen und Naturschutz. PhD Dissertation. Karlsruhe (TH): Universität Karlsruhe (TH).
17. Callahan JT (1984) Long-term ecological research. Bioscience 34:363-367
18. CBD (2012) Cities and biodiversity outlook: action and policy. UN Secretariat of the Convention on Biological. Montreal: Diversity.
19. Connelly M, Hodgson M (2013) Experimental investigation of the sound transmission of vegetated roofs. Appl Acoust 74:1136-1143
20. Cook-Patton SC, Bauerle TL (2012) Potential benefits of plant diversity on vegetated roofs: a literature review. J Environmen Manag 106:85-92
21. Darius F, Drepper J (1983) Ökologische Untersuchungen auf bewachsenen Kiesdaechern in West-Berlin. TU-Berlin: Diplomarbeit.
22. Denardo JC, Jarrett AR, Manbeck HB, Beattie DJ, Berghage RD (2005) Stormwater mitigation and surface temperature reduction by green roofs. Transact Asae 48:1491-1496
23. Dimoudi A, Nikolopoulou M (2003) Vegetation in the urban environment: microclimatic analysis and benefits. Energy Build 35:69-76
24. Dunnett N, Kingsbury N (2004) Planting green roofs and living walls. Portland, OR: Timber Press.
25. Dunnett N, Nagase A, Booth R, Grime P (2008) Influence of vegetation composition on runoff in two simulated green roof experiments. Urban Ecosystems 11:385-398
26. Dunnett N, Nagase A, Hallam A (2008) The dynamics of planted and colonising species on a green roof over six growing seasons 2001–2006: influence of substrate depth. Urban Ecosystems 11:373-384
27. Durhman AK, Rowe DB, Rugh CL (2007) Effect of substrate depth on initial growth, coverage, and survival of 25 succulent green roof plant taxa. Hort Sci 42:588-595

28. Dvorak B, Volder A (2010) Green roof vegetation for North American ecoregions: a literature review. Landsc Urban Plan 96:197-213
29. Ellenberg H (1986) Vegetation ecology of Central Europe. Cambridge: Cambridge University Press.
30. Emilsson T, Rolf K (2005) Comparison of establishment methods for extensive green roofs in southern Sweden. Urban For Urban Green 3:103-111
31. Emilsson T, Berndtsson JC, Mattsson JE, Rolf K (2007) Effect of using conventional and controlled release fertiliser on nutrient runoff from various vegetated roof systems. Ecol Eng 29:260-271
32. FBB (2012) Gründachmarkt von 2008 auf 2011 um etwa 19% gewachsen! [Online]. Saarbrücken, Germany: Fachvereinigung Bauwerksbegrünung e.V. (FBB). http://www.fbb.de/aktuelles/detailansicht/gruendachmarkt-von-2008-auf-2011-um-etwa-19-gewachsen-/53/ *webcite*. Accessed June 7 2012
33. Fernandez-Canero R, Gonzalez-Redondo P (2010) Green roofs as a habitat for birds: a review. J Anim Vet Adv 9:2041-2052
34. Forschungsgesellschaft Landschaftsentwicklung Landschaftsbau (FLL) (2008) Introduction to the FLL: guidelines for the planning, construction and maintenance of green roofing. 2008 edition of the green roofing guideline. Bonn: EV-FLL.
35. Francis RA, Lorimer J (2011) Urban reconciliation ecology: The potential of living roofs and walls. J Environ Manage 92:1429-1437
36. Franklin JF, Bledsoe CS, Callahan JT (1990) Contributions of the long-term ecological research-program - an expanded network of scientists, sites, and programs can provide crucial comparative analyses. Bioscience 40:509-523
37. Galtung J (1986) The green movement: a socio-historical exploration. Int Sociol 1:75-90
38. Gedge D, Kadas G (2004) Bugs, bees and spiders: green roof design for rare invertebrates. Greening Rooftops for Sustainable Communities. Portland, OR: Green Roofs for Healthy Cities.
39. Getter KL, Rowe DB (2007) Effect of substrate depth and planting season on Sedum plug survival on green roofs. J Environ Hortic 25:95-99

40. Getter KL, Rowe DB (2009) Substrate depth influences Sedum plant community on a green roof. Hortscience 44:401-407
41. Getter KL, Rowe DB, Robertson GP, Cregg BM, Andresen JA (2009) Carbon sequestration potential of extensive green roofs. Environ Sci Technol 43:7564-7570
42. Grant G (2006) Green roofs and facades. Bracknell: UK, IHS BRE Press.
43. Grant G (2012) Ecosystem services come to town: greening cities by working with nature. Oxford: UK, John Wiley & Sons, Ltd.
44. Grant G, Engleback L, Nicholson B (2003) English Nature Research Reports. In: Engleback E, Lia WS (eds) Green roofs: their existing status and potential for conserving biodiversity in urban areas, Peterborough: English Nature.
45. Green Roof Systems Project (2013) Marie Curie IAPP Green Roof Systems Project. University of Sheffield. Available: http://www.green-roof-systems.co.uk *webcite*. Accessed Aug 1 2013
46. Handreck K, Black N (2010) Growing media for ornamental plants and turf. Sydney: Australia, University of New South Wales Press Ltd.
47. Hodgson JG (1990) The role of autecological accounts. In: Hillier SH, Walton DWH, Wells DW (eds) Calcareous grasslands: ecology and management, Huntingdon: Bluntisham Books.
48. Jauch M, Fischer P (2000) Substrate im Dauertest. Deutscher Gartenbau, DeGa Spezial "GaLaBau Dachbegrünung". Stuttgart: Verlag Eugen Ulmer.
49. Kadas G (2011) Green roofs and biodiversity: can green roofs provide habitat for invertebrates in an urban environment?. London: LAP LAMBERT Academic Publishing.
50. Kent M, Coker P (1994) Vegetation description and analysis: a practical approach. Chichester: Wiley.
51. Köhler M, Poll PH (2010) Long-term performance of selected old Berlin greenroofs in comparison to younger extensive greenroofs in Berlin. Ecol Eng 36:722-729
52. Kolb W, Schwarz T (1999) Dachbegrünung: intensiv und extensiv. Stuttgart: Eugen Ulmer.
53. Kolb W, Schwarz T, Trunk R (1983) Zur Begrünung von Kiesdächern. Zeitschrift für Vegetationstechnik 6:143-151

54. Krupka BW (1992) Dachbegrünung Pflanzen- und Vegetationsanwendung an Bauwerken. Stuttgart: Verlag Eugen Ulmer.
55. Krupka BW (2006) Einschichtige Extensivbegrünungen auf Flachdächern ohne Gefälle. Berlin: Dach + Grün.
56. Liesecke HJ (1995) Wasserrückhaltung und Abflussbeiwerte bei extensiven Dachbegrünungen. Stadt und Grün 10:683-687
57. Liesecke HJ (1998) Langzeitentwicklung einer extensiven Dachbegrünung: Untersuchungen zum Substratverhalten und zur Vegetationsentwicklung eines 1985 ausgeführten Objektes. Stadt-und-Grün 47:428-436
58. Liesecke HJ (2006) Langzeitentwicklung von Substraten für die extensive Dachbegrünung. Dach + Grün 3:4-7
59. Likens GE (1989) Long-term studies in ecology: approaches and alternatives. New York: Springer Verlag.
60. Lundholm J, MacIvor JS, MacDougall Z, Ranalli M (2010) Plant species and functional group combinations affect green roof ecosystem functions. PLOS ONE 5(3):e9677
61. Mann G (1996) Die Rolle begrünter Dächer in der Stadtökologie. Biologie in unserer Zeit 5:292-299
62. McGuire KL, Payne SG, Palmer MI, Gillikin CM, Keefe D, Kim SJ, Gedallovich SM, Discenza J, Rangamannar R, Koshner JA, Massmann AL, Orazi G, Essene A, Leff JW, Fierer N (2013) Digging the New York city skyline: soil fungal communities in green roofs and city parks. PLOS ONE 8(3):e58020
63. Mentens J, Raes D, Hermy M (2006) Green roofs as a tool for solving the rainwater runoff problem in the urbanized 21st century? Landsc Urban Plan 77:217-226
64. Millennium Ecosystem Assessment (2005) Ecosystems and human well-being: biodiversity synthesis. Washington, DC: World Resources Institute.
65. Millennium Ecosystem Assessment (2005) Ecosystems and human well-being: synthesis. Washington, DC: World Resources Institute.
66. Miller C (2003) Moisture management in green roofs. Greening Rooftops for Sustainable Communities. Chicago, IL: Green Roofs for Healthy Cities.

67. Minke G, Witter G (1983) Häuser mit grünem Pelz: ein Handbuch zur Hausbegrünung. Fricke Verlag: Frankfurt/Main.
68. Moran A, Hunt B, Jennings G (2004) A North Carolina field study to evaluate greenroof runoff quantity, runoff quality, and plant growth. Greening Rooftops for Sustainable Communities. Portland, OR: Green Roofs for Healthy Cities.
69. Mueller-Dombois D, Ellenberg H (1974) Aims and methods of vegetation ecology. USA: Wiley.
70. Nagase A, Dunnett N (2010) Drought tolerance in different vegetation types for extensive green roofs: Effects of watering and diversity. Landsc Urban Plan 97:318-327
71. Oberndorfer E, Lundholm J, Bass B, Coffman RR, Doshi H, Dunnett N, Gaffin S, Kohler M, Liu KKY, Rowe B (2007) Green roofs as urban ecosystems: ecological structures, functions, and services. Bioscience 57:823-833
72. Osmundson T (1999) Roof gardens: history, design, and construction. London: W.W. Norton & Company.
73. Peck SW (2012) Green roof industry begins era of triple digit growth!. Toronto: Living Architecture Monitor. Green Roofs for Healthy Cities.
74. Poll PH (2008) Ökologie alter Dachbegrünungen - Ein Vergleich zwischen einfachen Jahrhundertwendedächern und komplexen Systemdächern der 80er Jahre. Diplom/Master's Diplomarbeit: Freien Universität Berlin, Berlin.
75. Roth-Kleyer S (2002) Was Dachsubstrate leisten müssen. Jahrbuch Dachbegrünung 2002. Bad Honnef: Bundesverband Garten-Landschafts- und Sportplatzbau.
76. Roth-Kleyer S (2006) Probleme beim Aufbringen von Substraten durch Gebläse, Teil 1. Neue Landschaft 51:27-30
77. Rowe D, Getter KL, Durhman AK (2012) Effect of green roof media depth on Crassulacean plant succession over seven years. Landsc Urban Plan 104:310-319
78. Sailor DJ (2008) A green roof model for building energy simulation programs. Energy and Buildings 40:1466-1478
79. Santamouris M, Pavlou C, Doukas P, Mihalakakou G, Synnefa A, Hatzibiros A, Patargias P (2007) Investigating and analysing the energy and environmental performance of an experimental

green roof system installed in a nursery school building in Athens, Greece. Energy 32:1781-1788

80. Schrader S, Boening M (2006) Soil formation on green roofs and its contribution to urban biodiversity with emphasis on Collembolans. Pedobiologia 50:347-356
81. Simmons MT, Gardiner B, Windhager S, Tinsley J (2008) Green roofs are not created equal: the hydrologic and thermal performance of six different extensive green roofs and reflective and non-reflective roofs in a sub-tropical climate. Urban Ecosystems 11:339-348
82. Snodgrass EC, McIntyre L (2010) The green roof manual: a professional guide to design, installation and maintenance. Portland, OR: Timber Press, Inc.
83. Snodgrass EC, Snodgrass LL (2006) Green roof plants: a resource and planting guide. Portland: OR, Timber Press, Inc.
84. Speak AF, Rothwell JJ, Lindley SJ, Smith CL (2012) Urban particulate pollution reduction by four species of green roof vegetation in a UK city. Atmos Environ 61:283-293
85. Stovin V, Vesuviano G, Kasmin H (2012) The hydrological performance of a green roof test bed under UK climatic conditions. J Hydrol 414–415:148-161
86. Takebayashi H, Moriyama M (2007) Surface heat budget on green roof and high reflection roof for mitigation of urban heat island. Build Environ 42:2971-2979
87. Thuring CE, Berghage RD, Beattie DJ (2010) Green Roof Plant Responses to Different Substrate Types and Depths under Various Drought Conditions. Horttechnology 20:395-401
88. Tonietto R, Fant J, Ascher J, Ellis K, Larkin D (2011) A comparison of bee communities of Chicago green roofs, parks and prairies. Landsc Urban Plan 103:102-108
89. Uhl M, Schiedt L, Mann G, Henneberg M (2003) Langzeitstudie zum Abflussverhalten begrünter Dächer. Parey Buchverlag, Berlin: Wasser & Boden.
90. UNFPA (2007) State of World Population 2007: Unleashing the Potential of Urban Growth. USA: United Nations Population Fund.

91. UNFPA (2011) The state of world population 2011. USA: United Nations Population Fund.
92. van der Maarel E (2005) Vegetation ecology. Oxford: Blackwell.
93. VanWoert ND, Rowe DB, Andresen JA, Rugh CL, Fernandez RT, Xiao L (2005) Green roof stormwater retention: Effects of roof surface, slope, and media depth. J Environ Qual 34:1036-1044
94. Vesuviano G (2013) A two-stage runoff detention model for a green roof. PhD thesis. Sheffield: The University of Sheffield.
95. Villarreal EL, Bengtsson L (2005) Response of a Sedum green-roof to individual rain events. Ecol Eng 25:1-7
96. Weiler SK, Scholz-Barth K (2009) Green roof systems: a guide to the planning, design, and construction of landscapes over structure. Hoboken, NJ: John Wiley & Sons.
97. Williams NSG, Rayner JP, Raynor KJ (2010) Green roofs for a wide brown land: Opportunities and barriers for rooftop greening in Australia. Urban For Urban Green 9:245-251
98. Wong NH, Chen Y, Ong CL, Sia A (2003) Investigation of thermal benefits of rooftop garden in the tropical environment. Build Environ 38:261-270
99. Yio MHN, Stovin V, Werdin J, Vesuviano G (2013) Experimental analysis of green roof substrate detention characteristics. Water Sci Technol 68(7):1477-1486

Chapter 2

# Invasive Species in Europe: Ecology, Status, and Policy

Reuben P Keller[1], Juergen Geist[2], Jonathan M Jeschke[3], and Ingolf Kühn[4]

[1]Program on the Global Environment, University of Chicago, Chicago IL 60637, USA

[2]Aquatic Systems Biology Unit, Department of Ecology and Ecosystem Management, Technische Universität München, Mühlenweg 22, 85354 Freising, Germany

[3]Department of Biology II, Ludwig-Maximilians-University Munich, Grosshaderner Str. 2, 82152 Planegg-Martinsried, Germany

[4]UFZ, Helmholtz Centre for Environmental Research--UFZ, Department of Community Ecology, Theodor-Lieser-Str. 4, 06120 Halle, Germany

## ABSTRACT

Globalization of trade and travel has facilitated the spread of non-native species across the earth. A proportion of these species become established and cause serious environmental, economic, and human health impacts. These species are referred to as *invasive*, and are now recognized as one of the major drivers of biodiversity change across the globe. As a long-time centre for trade, Europe has seen the introduction and subsequent establishment of at least several thousand non-native species. These range in taxonomy from viruses and bacteria to fungi, plants, and animals. Although invasive species cause major negative impacts across all regions of Europe, they also offer scientists the opportunity to develop and test theory about how species enter and leave communities, how non-native and native species interact with each other, and how different types of species affect ecosystem functions. For these reasons, there has been recent growth in the field of invasion biology as scientists work to understand the process of invasion, the changes that invasive species cause to their recipient ecosystems, and the ways that the problems of invasive species can be reduced. This review covers the process and drivers of species invasions in Europe, the socio-economic factors that make some regions particularly strongly invaded, and the ecological factors that make some species particularly invasive. We describe the impacts of invasive species in Europe, the difficulties involved in reducing these impacts, and explain the policy options currently being considered. We outline the reasons that invasive species create unique policy challenges, and suggest some rules of thumb for designing and implementing management programs. If new management programs are not enacted in Europe, it is inevitable that more invasive species will arrive, and that the total economic, environmental, and human health impacts from these species will continue to grow.

# INTRODUCTION

Globalization has integrated widely dispersed human communities into a worldwide economy. This process provides many benefits through the movement of people and goods, but also leads to the intentional and unintentional transfer of organisms among ecosystems that were previously separate [1]. Some of these species become established beyond their native range, a subset of these spread, and some of these have negative impacts and are termed invasive [2]. Although not all species moved beyond their native range become established [3], the large number of species transported and the range of pathways that move species mean that non-native species are now recognized as one of the major drivers of global biodiversity loss. They also cause significant damage to economies and human health [4-7].

Europe has been a centre for international trade for many centuries and has consequently seen the establishment of a large number of species. Some of these species have positive effects, including a subset of those introduced to enhance fisheries. Many other species, however, cause large negative impacts. These species cover a broad taxonomic range - from viruses and bacteria to fungi, plants, and animals - and affect all European nations and regions [6]. Where timelines are available, the number of non-native species established in Europe is generally increasing exponentially in freshwater [8] and terrestrial ecosystems [5,8-11]. This pattern is consistent with exponential increases in trade and travel [5]. Without increased efforts to manage pathways of introduction, the number of invasive species will continue to grow. Indeed, because there is often a significant lag period between species introduction and spread, it is likely that many future invasions have already been set in motion [12,13]. Consequently, the task of designing policies to reduce the transport and release of non-native species, and to manage those already established, has become a large priority for both national governments within Europe, and for the European Union [14,15].

In contrast to their negative impacts, non-native species offer large opportunities for ecologists to test fundamental theory [16, 17]. In many cases, established non-native species spread rapidly into new ecosystems. These species interact with native species through competition, predation, herbivory, introduction of disease, and resource use. When carefully studied, the effects of these perturbations allow insight into basic ecological processes. For example, community ecology seeks to understand why certain species communities exist (or do not), and how species within those communities interact. Because invasions involve new species entering communities and interacting with resident species, they offer insight into these patterns and processes. The very real need to better understand the process of invasion so that native habitats can be conserved, combined with opportunities to test fundamental ecological processes, has led to huge recent growth in the field of invasion biology (e.g. [18]), including in Europe.

The goal of this paper is to review the ecology and current status of non-native species in Europe, and to outline a framework for policy to reduce their number and impacts. We begin with a description of the *invasion process* through which non-native species are introduced, become established, and may then go on to have negative impacts. This structures the sections that follow which cover the pathways of species introduction, the characteristics of habitats and species that lead to establishment, the number of non-native species in Europe, and their impacts. Finally, we describe some characteristics of species invasions that create unique challenges compared to other environmental issues. We propose some general principles to govern the policy response to the risks from invasive species. Throughout the paper, we discriminate among three groups of organisms: terrestrial animals, terrestrial plants, and aquatic organisms. This distinction corresponds to the published literature, which in many cases is quite divergent among the groups. Our sections about aquatic organisms include both marine and freshwater species. We primarily review material specific to species invasions in Europe, but include work conducted elsewhere as necessary.

# THE INVASION PROCESS

To become invasive, a species must pass through a number of transitions (e.g. [2,19,20]). First, it must be entrained in a *pathway* and survive transit, where a pathway is a human-mediated process that facilitates the movement of species from one region to another. When it arrives in a region beyond its native range due to direct or indirect human intervention, it is referred to as*introduced*. Species that are not able to maintain self-sustaining populations, but that are occasionally found beyond cultivation, are often termed *casual* species. Next, if a species survives, escapes, and begins reproducing without direct human intervention it is referred to as *established*. Finally, we refer to a species as *invasive* if it spreads widely and causes measurable environmental, economic, or human health impacts.

Although the process described above is common to all invasions, different terminologies have been proposed, and these are often associated with different taxa or regions. For example, botanists may speak of an *invasive* plant as one that exceeds some predetermined rates of spread, whether or not it causes negative impacts (e.g. [19]). While we acknowledge the range of terminologies available, throughout this paper we use the definitions given in the previous paragraph.

The proportion of introduced species that become established can be quite low, as can the proportion of established species that spread and become invasive. These proportions vary with the taxonomy of the species in question, and the regions to which they are introduced. The 'tens rule' was proposed by Williamson [3] as a rule of thumb to approximate the proportion of species that make it through each step in the invasion process. This rule holds that approximately 10% of introduced species will become established, and approximately 10% of those species will become invasive. Hence, if 100 alien species were introduced, the tens rule holds that one will become invasive. The tens rule was developed with a focus on terrestrial plants, but has been applied by other authors to a wide range of species, often without thoroughly evaluating its validity. For example, recent studies show that for many animal taxa, the percentage of introduced species that become established and the percentage of established species that become invasive can exceed 50% [5,21]. According to the definition of invasive given above, Europe now contains >100 terrestrial vertebrates, >600 terrestrial invertebrates, >300 terrestrial plants, and >300 aquatic species that have become invasive [7].

The following sections are organized to follow the invasion process. In each, we describe the current state of knowledge about non-native terrestrial animals, plants, and aquatic organisms in Europe. We discuss the dominant pathways of introduction and the characteristics of ecosystems and species that most often lead to establishment. Next, we give estimates of the number of established non-native species from each group in Europe, and describe their impacts.

# INTRODUCTION PATHWAYS

There is a long history of classifying the pathways of non-native species. One of the oldest classifications was developed by Thellung [22, 23] for Central Europe. More recent schemes are from [9] and [24], and reviews and syntheses are from [25] and [11]. Hulme et al. [9], the most recent classification from a European perspective, identified three general mechanisms through which non-native species may enter a new region: importation as or with a commodity, arrival with a transport vector, and dispersal by the species themselves, either along infrastructure corridors (e.g. roads, canals) or unaided. The first of these, transportation as or with a commodity, arises from direct human movement of goods. Transportation *as* a commodity occurs when people identify a species as having desirable qualities and intentionally move it beyond its native range. Species introduced as commodities can be released intentionally (e.g. stocking animals to create populations for fishing and hunting, introduction of species as biological control agents) or can escape unintentionally (e.g. ornamental plants reproduce and spread beyond gardens, fish in aquaculture facilities escape). Species transportation *with* a commodity occurs when traded organisms arrive contaminated by non-native species, including diseases and parasites. These contaminants are not introduced intentionally, but may escape to become established and invasive. For example, the crayfish plague disease (*Aphamomyces astaci*) was introduced to Europe on North American crayfish imported for aquaculture. This disease has escaped, become established, and now infests and endangers native crayfish populations across Europe [26]. Another example is the Asian tiger mosquito (*Aedes albopictus*), which is native to South-east Asia but has spread to at least 28 countries, including several European countries, on ships as a contaminant of the trade in used car tires [27].

The second mechanism, arrival with a transport vector, refers to species that hitchhike on human modes of transport to reach regions beyond their native range. Modes of transport include ships (e.g. species in ballast water or attached to hulls), airplanes (e.g. contaminants of cargo, diseases/pests on food carried by passengers), and automobiles (e.g. as seeds caught in mud on tires). The influence of ships has been particularly strong in aquatic ecosystems, with many marine and freshwater species having become harmful invaders after dispersal in or on ships [28]. In Europe, this has included the spread of zebra mussel (*Dreissena polymorpha*) from the Ponto-Caspian Basin to the Atlantic and U.K. through the river and canal network.

The third mechanism is dispersal by the species themselves, either along infrastructure corridors or unaided. Corridor dispersal occurs when organisms move along canals, railways, roads, and other linear habitats created by humans. Examples include the introduction of species from the Red Sea into the Mediterranean Sea through the Suez Canal [29]. Unaided dispersal occurs when a non-native species becomes established in a neighbouring or nearby ecosystem, and then spreads without human intervention. An example is the continuing spread of invasive horse-chestnut leafminer moth (*Camariella ohridella*) [30] across Europe. Although this species was only introduced into a limited area, it has developed large populations and spread widely. Table 1 lists a number of important pathways that have transported species that are now established to Europe.

**Table 1:** Some important pathways of introduction for non-native terrestrial animals, terrestrial plants, and aquatic organisms

| Group | Sub-group | Dominant pathways | References |
|---|---|---|---|
| Terrestrial vertebrates | Mammals | Intentional introduction as commodity (for hunting, 'fauna improvement', fur farming, as pets, or for zoos), then either intentional release or accidental escape | [9,31] |
| | Birds | Intentional introduction as commodity (for hunting, 'fauna improvement', as pets, or for zoos or bird parks), then either intentional release or accidental escape | [9,32] |
| | Reptiles/ amphibians | Intentional introduction as commodity (for 'fauna improvement', as pets, food source, or biological control agents), then either intentional release or accidental escape | [9,32] |
| Terrestrial invertebrates | Insects | Unintentional introduction as contaminants or stowaways, sometimes deliberate release as biological control agents | [9,33] |
| | Other | Unintentional introduction as contaminants or stowaways | [33] |

| | | | |
|---|---|---|---|
| Terrestrial plants | Vascular plants, mosses, and lichens | Intentional introduction as commodity for garden trade (ornamentals), horticulture, unintentional introduction as contaminant of plants introduced for agriculture and ornamental trade (e.g., soil contaminants in plant pots) | [9,10,34,126] |
| Aquatic organisms | Fishes | Intentional introduction for aquaculture, stocking to improve recreational and commercial fisheries (including illegal stocking), as well as for weed and mosquito control, unintentional introductions with ship ballast water, ornamental species, fishing bait releases | [75,127] |
| | Crustaceans | Intentional introduction for aquaculture, ornamental reasons (Decapoda), unintentional introductions with ship ballast water, canals | [38,128-130] |
| | Mollusks | Unintentional introductions with shipping, waterways, accidental (e.g. during fish stocking), but also from garden pond and aquarium trade | [131,132] |
| | Plants | Intentional introduction for ornamental (aquarium and watergarden) trade, often further spread by boats and waterbirds | [133,134] |

Pathways of introduction for organisms established in Europe. Pathway lists given are not comprehensive and were chosen to give an indication of the total range of vectors, not necessarily those that are most important for each group.

Keller et al. Environmental Sciences Europe 2011 23:23, doi:10.1186/2190-4715-23-23

## Terrestrial Animals

Most terrestrial vertebrate animals established in Europe (or elsewhere throughout the world) were intentionally introduced as commodities, e.g. by the pet trade, the live food trade, or as stock for the trade in fur pelts [9, 31-33]. Although some of these pathways have been modified and restricted to reduce the risk of invasion, many remain very active. For example, the pet trade remains a dominant pathway for the introduction of new invasive species to Europe [31, 32].

Many other terrestrial animal species, especially invertebrates, have been introduced across Europe unintentionally, mostly as stowaways or contaminants on traded products, on vehicles (e.g. ships), or as diseases/parasites of plants, animals, and humans [9,31-33]. In general, much less is known about introductions of species that were introduced unintentionally because they are usually not recorded until they become established.

## Terrestrial Plants

Almost two thirds (62.8%) of the established plant species in Europe were introduced intentionally for ornamental, horticultural, or agricultural purposes. The remaining species were introduced unintentionally, mostly associated with transport vectors, or as contaminants of seeds and other commodities [34]. Of the terrestrial plant species that have escaped from human cultivation, some were intentionally released (i.e. planted in the wild to 'improve' the landscape), some were contaminants or stowaways, and only a few arrived unaided [9]. Consistent with increases in international trade, there has been a steady increase in the number of established non-native terrestrial plant species discovered in Europe, especially since 1800. Currently, an average of 6.2 species not native to any part of Europe is newly recorded as established each year. An average of 5.3 European species are found in parts of the continent outside their native range each year[10].

Much less is known about the introduction and spread of non-native lower plants and fungi, and about changes in the number of non-native species in Europe over time. It is known that these taxa can have enormous impacts, with perhaps the most damaging examples being diseases of crops and livestock. The potato famines that occurred across Western Europe during the nineteenth century, for example, were caused by the invasive potato blight (*Phytophthora infestans*) [35], introduced from North America.

## Aquatic Organisms

The pathways of introduction for aquatic organisms are generally known with less precision than those for terrestrial organisms. This arises partly because many aquatic species are introduced unintentionally with few (if any) records kept. Additionally, the difficulty of sampling in marine and freshwater environments means that a species may be well established, and may have spread from its initial site of introduction, before it is recorded.

Shipping has been by far the dominant pathway for the introduction of non-native marine species to the European Atlantic coast (47% of established non-native species) and to the Baltic Sea (45%,[36]). This pathway has also been a significant factor in freshwater animal introductions to Europe (25% of established non-native species, [37]). The shipping network creates connections among aquatic ecosystems across the globe, and organisms are frequently transported in the ballast water of ships, or attached to hulls as fouling organisms [28]. Ballast water is taken on to increase a vessel's weight when it is not fully laden with cargo. As this water is taken on, any organisms in the water are also sucked in. Vessels then travel to subsequent ports, and surviving organisms can be discharged with ballast water if the vessel takes on more cargo.

The opening of canals that link previously isolated water bodies has created many opportunities for the introduction and spread of non-native species. In the Mediterranean Sea, 54% of established non-native species arrived by dispersing through the Suez Canal [36]. Canals have also had a profound impact on the establishment and spread of non-native freshwater species in Europe, and this impact is tightly linked to shipping. There are now river and canal connections running from the Black Sea across Europe to the mouth of the Rhine River, and north to the Baltic Sea [38]. These connections have served as invasion corridors for many species native to the Ponto-Caspian into Western and Northern Europe. It is estimated that 8% of non-native freshwater animal species in Europe arrived by using natural dispersal mechanisms to move through canals [37]. In addition, many of the species that arrived through shipping (see previous paragraph) could only do so because of the existence of canals.

Despite the influence of shipping and canals, the most important pathways for the introduction of non-native freshwater animal species to Europe have been stocking (30% of species) and aquaculture (27%) [37]. Stocking has been largely of fish to create new wild populations, while aquaculture introductions have arisen from the unintended escape of farmed species and their associated organisms (e.g. parasites). Aquaculture has also been important for introductions of marine species to the Atlantic coast, Baltic Sea, and Mediterranean, accounting for 24%, 18%, and 11% of established species, respectively [36].

The final pathways mentioned here are the trades in ornamental (predominantly aquarium and watergarden) and aquaculture species. Ornamental introductions have been especially important in freshwater ecosystems, accounting for 8% of established non-native animal species. Ornamental introductions also appear to be by far the dominant pathway for introduction of aquatic plants. For example, in Great Britain 22 of 31 established non-native freshwater plant species were introduced for the ornamental trade [8]. The aquaculture trade has unintentionally introduced a large number of non-native aquatic species as contaminants of intentionally introduced species such as fish or shellfish. This is true for both marine and freshwater habitats. For example, the unintentional introduction and spread of the brown algae *Sargassum muticum*, the Japanease kelp *Undaria pinnatifida*, and the snail *Ocinebrellus inornatus*, as well as the oyster parasites *Mytilicola orientalis*and *Myicola ostreae*, all occurred because these species inadvertently arrived associated with marine shellfish imported from Asia to Europe for aquaculture [36].

# CHARACTERISTICS OF HIGHLY INVADED REGIONS

The number of invasive species found in a region depends on the number of species that have been introduced, the proportion of introduced species that have become established, and the proportion of established species that have gone on to cause impacts. When investigating differences among regions, invasion biologists have generally left aside the pathways and process of introduction and focused instead on the proportions of introduced species that become established, and of established species that become invasive [39]. Ecosystems where these proportions are high have been assumed to be highly invasible, while others have been deemed relatively resistant.

Different theories have been proposed to explain why some regions appear more invasible than others. Perhaps the most influential has been the *biotic resistance* hypothesis, an early champion of which was Charles Elton, often referred to as the scientist who founded the field of invasion biology (e.g. [19]; although we note that Charles Darwin [40] and other biologists had already written about the spread and impacts of non-native species, [11,41]). This hypothesis holds that regions with high biodiversity and a relatively low level of disturbance, especially disturbance from humans, are more resistant to establishment by non-native species [42]. The rationale is that less diverse and/or more disturbed ecosystems are likely to have more vacant niches that introduced species can inhabit. Although this is an intuitively appealing argument, there has been little empirical evidence generated to support it. In fact, especially at larger spatial scales, there is increasing evidence that highly diverse habitats are actually more prone to non-native species establishment (e.g. [43-45]. Several authors have attempted to reconcile the contrasting theory and evidence, but no consensus has yet been reached (e.g. [46, 47]). A large obstacle to finding this consensus comes from the difficulty of quantitatively assessing levels of disturbance and the presence of vacant niches.

As well as trying to reconcile theory and observed patterns in species establishment, ecologists are now paying more attention to the introduction process. Recent results show that the number of species introduced to a region may be at least as important as the invasibility of the region in determining how many species become established [48-52]. This is discussed further in the following sections.

## Terrestrial Animals

There is a particular lack of support for the biotic resistance hypothesis when terrestrial animals are considered (see [52] and references therein). It has become clear in recent years that the key difference among regions with different numbers of established terrestrial animals is the number of species that have been introduced. But which regions are the ones that receive more introductions than others? The answer is that regions with high human impact typically receive more species introductions than other regions, and that this leads to them containing more established species. For example, 12 non-native mammal species have established in France, nine in Germany, but just two in Portugal. Contrary to what would be predicted from the biotic resistance hypothesis, it is not easier for introduced mammals to establish in countries with high human impact, but these countries host more non-native mammals than other countries because they have received more species introductions [52]. Differences among the number of established birds in European countries can also be best explained by differences in number of introduced species [51].

## Plants

Patterns of plant invasion in Europe offer little support for the biotic resistance hypothesis (e.g. [44,53,54]). Instead, apart from broad habitat type, the number of species introduced and their propagule pressure appear to be the most important determinants of the number of established non-native species in any given region (e.g. [55,56]). The most invaded habitats in Europe are in heavily transformed landscapes such as agricultural land, coniferous forests, urban areas, and dump or construction sites [57]. In contrast, natural and semi-natural environments such as broad leaved and mixed forests, pastures, natural grasslands, moors, heathlands and peatbogs have remained relatively uninvaded [57]. This pattern is consistent with that observed for terrestrial animals - that sites experiencing high levels of human disturbance and high propagule pressure tend to be the most invaded. Disturbance increases plant invasions because it leads to loss of native species that could compete with introduced non-native species, and because it increases availability of resources [58]. High propagule pressure occurs in the same regions because human activities lead to many plant introductions [59].

The highest proportions of established terrestrial plant species in Europe occur in agricultural landscapes, especially in eastern Britain, northern France, Central and Eastern Europe, and the Po floodplain in Italy. In contrast to a global pattern of Mediterranean-type ecosystems being highly invaded, the European Mediterranean biogeographic region is relatively uninvaded, probably due to a long history of human presence and prehistoric introductions in the Mediterranean Basin, which may make its ecosystems relatively resistant against recently introduced species [57]. Additionally, the Mediterranean Basin acted as more of a donor than recipient region for species introductions during the colonization of the New World [60].

It has been argued that harsh environments (e.g. alpine habitats) might not be suitable for non-native species. However, these are also often the habitats that experience low propagule pressure [61]. Hence, it is clear that the intensity of human activities that increase or facilitate propagule pressure, pathways of introduction, intensity of disturbance, and eutrophication, are important determinants of non-native plant invasions. For many taxa in Europe, it is even more important than climate or other features of the physical environment [62].

## Aquatic Organisms

European aquatic ecosystems containing the highest numbers of non-native species tend to be those with high connectivity to other ecosystems, high frequency of human access (e.g. for transportation or recreation), and high disturbance. These include boat harbours, recreational areas at lakes (jetties etc.), and the many canals that now cross Europe. More remote water bodies, including mountain lakes and headwater streams, tend to be least and last invaded. Thus, propagule pressure can largely explain the intensity and diversity of established non-native species in aquatic environments [3, 20]. In marine ecosystems, the number and frequency of pathways, tidal movements, availability of empty niches, and availability of different substrate types for settlement are the main factors that determine susceptibility to invasion, with highest rates of non-native species establishment typically found in shallow coastal zones [36]. Consequently, marine ecosystems with high numbers of established species in Europe include the eastern Mediterranean with hundreds of introductions through the Suez Canal [63], as well as the Gulf of Finland, the Gulf of Riga, the coastal lagoons [64-66], and the Oosterschelde estuary [67]. Of the 737 non-native multicellular animal species recorded from European seas, 569 have been found in the Mediterranean, 200 along the Atlantic coast (Norway to the Azores, including the UK and Ireland), and 62 in the Baltic Sea [36]. Numbers in the Mediterranean are highest because of the Suez Canal, the role of the Mediterranean as a long-time hub of international shipping, and a surge in development of mariculture [36].

# CHARACTERISTICS OF INVASIVE SPECIES

An alternative perspective comes from asking whether there are traits of non-native species that are associated with successful passage through the invasion process. Ecologists have been asking this question for several decades (e.g. [68]), often concentrating on intuitive life-history traits such as early reproduction, high reproductive output, or a non-specialized diet (reviewed by [69,70]). This work has recently become more important because many nations, including several in Europe and the European Union, have begun to develop risk assessment programs for non-native species[71-75]. The development of risk assessment tools begins with the search for patterns in species traits that are associated with successful passage through the invasion process. If robust patterns are identified, they can be applied to non-native species to determine the likelihood that they will become established, spread, and/or become invasive (see [2,76,77] for reviews). In this way, knowledge of non-native species traits can support pro-active efforts to prevent new invasions.

Although the search for traits of invasive species has been fruitful, recent results have shown that propagule pressure can be confounded with invasiveness [39]. As for the discussion above that focused on differences in invasibility among regions, invasion biologists have traditionally left aside introduction and focused on establishment and spread when looking for differences between the traits of invasive and non-invasive species. Recent studies have challenged this by showing that those species most likely to establish are often those introduced in the highest numbers and most frequently. This does not mean that the search for traits of invasive species is not worthwhile, but does indicate that additional factors are important.

## Terrestrial Animals

Recent studies of the characteristics of invasive terrestrial animal species have shown that invasive species tend to have been introduced in higher numbers and more frequently than non-invasive species [69, 78]. Mammals and birds that are hunted by humans are more frequently invasive than other species of mammals and birds because they have been more frequently introduced than other species, even though their establishment success is not higher than that of other species [69]. The same is true for mammals and birds with large native ranges which also become invasive more frequently than species with smaller native ranges. Their establishment success has not been shown to be higher than that of non-native species with smaller native ranges, meaning that this pattern is best explained by their increased frequency of introduction[69,79].

Terrestrial animals living in association with humans tend to become invasive more often than other species [69]. Good examples are the Norway rat (*Rattus norvegicus*) which reaches extremely high population densities in cities; the rose-ringed parakeet (*Psittacula krameri*), native to Africa and Asia and also often very abundant in human settlements; and the harlequin ladybird (*Harmonia axyridis*), native to Asia and infamous for its large aggregations in buildings during winter [6]. Thus, a clear understanding of human activities, both in terms of propagule pressure and the location of human settlements, is very important for understanding patterns of establishment, spread, and harm for non-native terrestrial animals.

There are also species-level biological traits linked to terrestrial animal invasiveness. For example, behavioural flexibility as expressed by brain size was among the best predictors of invasiveness in a study of non-native birds [80]. Mammals and birds with high ecological flexibility (indicated by the number of different types of food they consume or the number of different types of habitat they use) also tend to be more invasive than other species [69, 81, 82]. Thus, species that are relatively more behaviourally and ecologically flexible tend to become invasive more often than other species.

## Terrestrial Plants

Several factors are related to the invasion success of individual plant species. First, non-native plants that have been introduced and/or planted more frequently (i.e. higher propagule pressure) are more likely to become established and to have a larger range (e.g. [83]). Second, residence time (i.e. time since introduction) in the non-native range is important, with those species that have been present for longer tending to have larger ranges [84]. This is an effect of having the opportunity to fulfil more life cycles and also simply having the time to spread further. The importance of residence time is also associated with propagule pressure, as species that were introduced a long time ago are likely to have been introduced many times since the first introduction. Third, species with larger native ranges are more likely to successfully establish beyond their native range. Similar to terrestrial animals, this is presumably associated with a higher probability that the species will be accidentally transported [85]. Additionally, species with a large native range are more likely to have a strong climate match to at least part of Europe, making them pre-adapted to survive there. Fourth, once terrestrial plant species have been introduced to the new range, traits of the species are important for determining whether they will successfully establish, spread, and cause harm. Traits known to promote passage through the invasion sequence include long flowering season, being an annual, vegetative spread, having multiple dispersal vectors [85], high maximum relative growth rate, and high resource allocation to shoots and leaves [86,87].

Many studies that have attempted to relate biological traits to invasiveness have explained little of the variation and have neglected trait interactions. Including interactions among traits (i.e. explicitly considering that one trait value might have a different influence on invasion success in the presence of other traits) can result in much better explanatory models. Küster et al. [88] found that trait interactions accounted for >40% of the variation that could explain invasion success of non-native terrestrial plants in Germany. Interestingly, long flowering season was beneficial for self-pollinated species, but was disadvantageous for wind pollinated species, and had no effect on insect-pollinated species. Furthermore, the effect of timing of the end of the flowering season on invasion success differed among plant species with different vegetative reproduction strategies or different levels of ploidy (number of chromosome sets in the cell). Thompson and Davis [89], however, argue that such analyses tell us very little because successful invaders do not differ in their traits from those of widespread native plant species. Despite this critique, incorporating statistical interactions among traits should increase our knowledge of characteristics that make a species likely to expand or contract its range, whether non-native or native.

## Aquatic Organisms

As for terrestrial animals and plants, there are some general rules that separate non-native aquatic species that successfully pass through the invasion sequence from those that do not. Some of the characteristics that influence invasion success are associated with biological traits whereas others are closely linked to interaction with humans. For example, species introduced intentionally (and hitchhiking species associated with them) because they have desirable attributes tend to be more successful than undesired species. Prominent examples in marine environments include the introductions of alien shellfish species (e.g. *Crassostrea gigas* introduced to France from Japan) for mariculture, which have arrived with several associated parasites and algae. Additionally, many of the most widespread non-native aquatic species in Europe are generalists that can tolerate a wide range of environmental conditions such as water temperature and salinity. European brackish water systems hold a great diversity of invaders which may be due to their poor native species richness [90] and the great ecological plasticity of the non-native species that have established. In addition to the breadth of ecological niches, similarity of environmental conditions in the donor and the receiving region can also be crucial [28]. For example, most of the 569 non-native species in the Mediterranean are thermophilic and originated from tropical waters in the Indo-Pacific, the Indian Ocean, the Red Sea, and pan-tropical regions [36].

Differences in life history and reproduction can differentiate between invaders and rare species. This is evident in freshwater unionid mussels, which are among the most critically imperiled freshwater taxa both in North America and Europe [91, 92]. These species produce glochidia larvae that need to attach to a suitable fish host to survive. The high degree of specialization and the complex life cycle of unionids probably contribute to the decline in this group. In contrast, invasive mussels of the genus *Dreissena* are less specialized and produce free veliger larvae, allowing for a higher rate of dispersal through passive transport (e.g. in the ballast water of ships). Non-native marine species, which have been predominantly introduced to Europe through shipping, are also more likely to have larval stages that are tolerant of conditions in ships.

Reproduction rates tend to be higher in invasive aquatic species compared to those in most non-invasive species (e.g. [93]). High reproduction can facilitate rapid spread and secondary introductions into other areas. Mode of feeding can also be important, with filter-feeding freshwater macroinvertebrates in Europe and North America known to be more successful at invading than predator macroinvertebrates [94]. This has the impact of enhancing energy flow between benthic (i.e. bottom) and pelagic (i.e. open water) regions because algae that are produced mostly in the pelagic zone are consumed by the benthic filter feeders. For intentional introductions of fishes, where large predator species tend to be most popular, competition and top-down regulation may be more important. Overall, effects of non-native species tend to be greater when they establish in high abundance and have strong functional distinctiveness from native species [95].

# NUMBER OF ESTABLISHED NON-NATIVE SPECIES IN EUROPE

At least several thousand non-native species are now established in Europe [6]. These include species not native to any part of Europe, as well as species native to one part but now established in another. The following sections give estimates of the number of species in different habitat categories that are established in Europe and not native to any part (unless stated otherwise). These figures should be seen as low estimates of the true numbers of established species because only recorded species are included; it is likely that many additional species are established but not yet recorded.

## Terrestrial Animals

According to the DAISIE database, there are 33 non-native established mammal species [31] and 77 established bird species in Europe [32]. These figures are probably quite accurate because these taxa are relatively large and easy to distinguish from native species. For the same reason, the estimate of 55 established reptiles and amphibians in Europe [32] is also probably quite accurate. In contrast, estimates for invertebrates are likely to be more severe underestimates because these species are more difficult to collect and identify. Within terrestrial invertebrates, data for insects tend to be more accurate than those for other invertebrates [33]. Insects are also the dominant group among non-native terrestrial invertebrates in Europe: of 1,522 established species, 1,306 (86%) are insects [33]. This high proportion is not unexpected, however, as 85% of the world's known invertebrates are insects [96].

## Terrestrial Plants

Terrestrial plants are generally well sampled, but it can be difficult to assess total numbers of established non-native species because the same species is often given different scientific names in different parts of Europe. According to the DAISIE database [10], 5,789 plant species have been recorded from the wild (not necessarily established) in at least one European country to which they are not native. These species come from 213 families and 1,567 genera, and include 2,843 species not native to any European country. A total of 3,749 plant species are known to be established in at least one European country to which they are not native, and 1,780 of these species are entirely non-native to Europe. We note that the numbers just given include all species recognized as non-native, irrespective of their date of introduction [97]. Traditionally, in those countries where records are available, botanists distinguish between species introduced before the European discovery of the Americas (1492) and those introduced later.

## Aquatic Organisms

It is estimated that 737 non-native multicellular animal species from marine environments and 262 non-native freshwater animal species are now established in Europe [36,37]. These comprise a wide range of taxa, including fishes, arthropods, mollusks, platyhelminthes, and annelids. For aquatic plants, it is estimated that at least 260 species not native to any part of Europe are established in inland waterways [34].

The number and diversity of non-native species is variable across different regions of Europe. For example, in Great Britain the 134 established non-native species in freshwater ecosystems are dominated by plants (31), fishes (18), non-decapod crustaceans (17), platyhelminths (15), and amphibians (11; full list in [8]). In Italy, the patterns are somewhat different, with the 112 non-native species from inland aquatic systems being dominated by fishes (38), non-decapod crustaceans (28), and gastropods (7; full list in [98]). In each case, it is reasonable to expect that non-native species from groups such as fishes and crayfishes are better represented in the data than records of species that are smaller and less often sampled (e.g. annelids).

Just seven non-native vascular plants have been identified in European marine ecosystems [34]. In contrast, numbers of non-native marine species (i.e. including animals and other multicellular organisms) are much larger. The three main marine biogeographic regions of Europe are the Mediterranean, the Atlantic coast, and the Baltic Sea; these contain 569, 200, and 62 established non-native species, respectively [36]. These species cover a large taxonomic range, from fishes to barnacles to plants.

# IMPACTS OF NON-NATIVE SPECIES IN EUROPE

Invasive species have a large and diverse range of impacts in Europe. This diversity of impacts is mainly driven by the diversity of species, and makes generalized statements about types of impact difficult. However, it is clear that invasive species have significant negative impacts on many native species and almost all ecosystems, on the European economy, and on human health (recently reviewed by [7]). Economic impacts alone are estimated to be at least 12.5 billion EUR per year, and are probably over 20 billion EUR [99].

## Terrestrial Animals

Ecological impacts of invasive terrestrial animal species include predation/herbivory, competition, transmission of diseases, and hybridization with native species. Economic effects include impacts on human infrastructure, human health, human social life, livestock, plant production, and forestry[7,100,101]. For example, Norway rats (*Rattus norvegicus*) predate many native species and have caused declines in native bird species and small mammal species. They are also a reservoir and vector of many diseases, including hepatitis E, leptospirosis, hantavirus, and Q fever. The invasive American mink (*Neovison vison*) is a competitor of the European mink (*Mustela lutreola*) which is now listed as endangered on the IUCN Red List of Threatened Species. Another example of an invasive competitor of a native species is the North American grey squirrel (*Sciurus carolinensis*, Figure 1A) that threatens the native red squirrel (*Sciurus vulgaris*), especially in the U.K. and Italy. The Canada goose (*Branta canadensis*, Figure 1B) is also an abundant invader. It hybridizes and competes with native geese, and its droppings can cause human health hazards and algal blooms. The Asian tiger mosquito (*Aedes albopictus*) competes with native mosquito species; its bites are a nuisance to humans, and it is a vector for diseases such as West Nile virus. Another invasive terrestrial invertebrate with severe impacts is the harlequin ladybird (*Harmonia axyridis*). Its tendency to overwinter in large aggregations inside buildings is a nuisance to many people, and the unpleasant odour of its body fluids can destroy the taste of wine. It also threatens native ladybirds and other European insect species [6]. Overall, invasive terrestrial invertebrate species cause costs of at least 1.5 billion EUR per year in Europe, and invasive terrestrial vertebrates cause costs of at least 4.8 billion EUR per year [99].

**Figure 1:** Examples of high-impact invasive species in Europe. (A) grey squirrel (*Sciurus carolinensis*), native to North America, ©Jeschke; (B) Canada

goose (*Branta canadensis*), also native to North America, © Jeschke; (C) giant hogweed (*Heracleum mantegazzianum*), native to the Caucasus region, © Denholm, NJ Dept of Agriculture, Bugwood.org; (D) round goby (*Neogobius melanostomus*), native to the Caspian, Black, and Azov Seas, © Aquatic Systems Biology Unit, TUM; (E) common ragweed (*Ambrosia artemisiifolia*), native to North America, © Bodner, Southern Weed Science Society, Bugwood.org; (F) signal crayfish (*Pacifastacus leniusculus*), native to North America, © Aquatic Systems Biology Unit, TUM.

## Terrestrial Plants

Many invasive plant species in Europe are primarily recognized as agricultural or forestry weeds. Additionally, 17 out of the 18 plant species recorded among the most damaging invasive species in Europe [6] are known to reduce the habitat of native species [34]. Eight of them are reported to disrupt community assemblages, for example by impacting plant pollinator networks [34]. Non-native plant species can also hybridize with closely related native species so that distinctive genotypes of native plants are lost [102]. Species such as Japanese knotweed (*Fallopia japonica*) and Himalayan balsam (*Impatiens glandulifera*) grow and are nuisance species along railway lines (the former) and waterways (both). Other plant species can cause severe health problems. For example, giant hogweed (*Herracleum mantegazzianum*, Figure 1C) produces sap that causes skin lesions to humans upon contact [103]. The pollen of invasive ragweed (*Ambrosia artemisiifolia*, Figure 1E) is highly allergenic to humans, and estimates of associated medical costs in Germany range between 17 and 47 million EUR annually [104]. According to Vilà et al. [7], the most costly plant invaders affecting nature conservation, agriculture, forestry, and fisheries in Europe are pigface species (*Carpobrotus* spp.). These produce annual costs for control and eradication in Spain of approximately 0.58 million EUR [105]. Overall, invasive terrestrial plants cause costs of at least 3.7 billion EUR per year in Europe [99].

## Aquatic Organisms

Invasive species are considered one of five major threats to aquatic biodiversity worldwide [4], with particularly large impacts on freshwater habitats [106,107]. The isolated nature of most freshwater habitats means that natural spread of aquatic organisms into new habitats occurs at low frequencies. In turn, this means that aquatic communities tend to be more different to each other, and thus that the increased rates of species movement caused by human pathways have large potential for impacts on biodiversity.

Traditionally, the study of invasions in aquatic ecosystems has had a strong focus on economically important and visible species, whereas invasive populations of small taxa (e.g. plankton) or groups that are difficult to identify at the species level (e.g. chironomids) have rarely been considered. As a larger and more visible species, North American signal crayfish (*Pacifastacus leniusculus*, Figure1F) have been relatively well monitored and recorded. They were introduced to Europe primarily for aquaculture, have spread rapidly, and are now considered one of the major threats to the indigenous crayfish fauna [108]. In addition to their competitive behaviour [109], North American crayfish are hosts of the crayfish plague (*Aphanomyces astaci*), an oomycete fungus that causes a lethal disease to European crayfish [110]. Also, the introduction of non-native salmonids and gobiids (e.g. *Neogobius melanostomus*, Figure 1D) has resulted in the decline and even extinction of indigenous species, and caused ecosystem shifts in lakes and streams [6,111]. In economic terms, the zebra mussel *Dreissena polymorpha*, which can completely block cooling systems in hydropower plants, probably has the greatest impact of all freshwater invaders.

In marine habitats, negative effects of non-native species include declines in native species richness and abundance. These impacts have been associated with the invasion of *Caulerpa taxifolia* into the Mediterranean [112], and with the high mortality rates of European oysters (*Ostrea edulis*) due to competition with introduced Pacific oysters (*Crassostrea gigas*) and damage from introduced parasites. Despite these examples, there is little comprehensive evidence for most impacts of invasive marine species, and there are some examples of economic benefits. For example, the release of the red king crab (*Paralithodes camtschaticus*) into the Barents Sea and its southward spread along the Norwegian coast has provided an additional fishery and income for fishermen in the order of 9 million EUR per year [36]. Nonetheless, negative impacts of invasive aquatic species in Europe are high and have been estimated to cost at least 2.2 billion EUR per year [99].

# MANAGEMENT AND POLICY FOR INVASIVE SPECIES

## Policy Challenges

Compared to other environmental problems, invasive species present at least six particular challenges. First, their impacts tend to increase over time as populations become larger and spread [113]. In contrast, other environmentally damaging activities, such as the release of chemical pollutants, generally decrease in severity over time after the activity is ended. This means that populations of invasive species can best be managed through rapid eradication of new populations [114]. This is most feasible soon after a species becomes established, but the type of biological information needed to support eradication and the resources and political will to eradicate a harmful invader are generally not available until the species has spread and become invasive. By this point there are fewer options to control the population, and the levels of resources required, and the possibility for undesirable side effects on other ecosystem components, are usually prohibitive.

Second, many other issues for environmental policy, such as forestry and mineral exploration, can be effectively managed with little concern for the policies followed by one's neighbours. In contrast, invasive species readily cross borders as long as there is suitable habitat on the other side [1]. This means that efforts to prevent invasions need to be mindful of the efforts being made in neighbouring jurisdictions [1]. As European economies become more integrated there are fewer barriers, such as border checks, to the spread of invasive species.

Third, invasive species often travel as contaminants of valuable trade. For example, most large ships carry ballast water to increase their weight when they are not fully loaded with cargo. This ballast is taken on at one port, along with any organisms in the water, and discharged at subsequent ports as cargo is loaded. Unintentional transfer of organisms in ballast tanks is now the largest single cause of marine invasions [115], and a strong factor in freshwater invasions[28,116]. Reducing the spread of invasive species via this pathway would either require constraints on where ships travel, or the installation onto all ships of expensive ballast treatment technology. Either of these options would increase the cost of shipped goods. Thus, control of invasive species often involves trade-offs with other activities, complicating decisions about how impacts can best be managed [114].

Fourth, it is usually impossible to determine the exact conditions that will lead to an invasion. Risky activities, such as collecting ballast water in one port and dumping it in another, may occur many times before an invader becomes established. The invasion of the Laurentian Great Lakes by zebra mussels (*Dreissena polymorpha*) in the late 1980s, for example, had been predicted for over a century before it actually occurred [117]. These predictions were based on shipping connections between the Great Lakes and regions where zebra mussel was established, and on the fact that the species is well suited to Great Lakes conditions. Why it took so long for the invasion to occur and why it occurred when it did remains unknown. This general uncertainty about invasions complicates management because economic activities with definite and well-quantified immediate benefits must be balanced against risks from the stochastic invasion process [118].

Fifth, controlling the spread of invasive species requires international cooperation, and this is often difficult to achieve [1]. For example, in response to the risk from ballast water invasions, the International Maritime Organization (IMO) produced the *International Convention for the Control and Management of Ships Ballast Water and Sediments* in February 2004. This convention requires the establishment of ballast water management systems on ships, with the goal of preventing the movement of live organisms. The dates at which each ship would be required to have ballast water treatment facilities differ based on ship size and age, but fall between 2009 and 2016. Despite this convention being finalized for several years and despite a widespread consensus that its implementation could reduce rates of invasion, it has not yet come into effect because too few countries have ratified it. Many other pathways of species invasion could be best managed at an international level, but policies have rarely been negotiated and implemented at this level [1].

Finally, there is often a considerable time lag, occasionally as much as several centuries, between the introduction of non-native species and their spread. These time lags appear to be especially long for terrestrial plants [12]. In a recent paper, Essl et al. [13] have shown that this may lead to an 'invasion debt' in which human activities many decades ago may have already set in process invasions that have not yet been recorded. In their paper, Essl et al. [13] showed that for several groups of plants and animals the socio-economic conditions in Europe during 1900 were more closely related to current levels of invasions than conditions in the year 2000, even though many species were introduced in the second half of the twentieth century. This means that current activities in Europe may be setting in motion many future invasions.

## Bioeconomic Approaches to Policy and Management

Ecological studies of species invasions have increased in number and sophistication over the last several decades [19]. This has created a much clearer picture of the number of species that are established beyond their native range, and the types and severity of impacts that they cause. Recognition of these impacts has led to increased calls for management and policy (e.g. [119]). Although many nations now have policies to reduce the impacts of invasive species, invaders continue to arrive at an increasing rate, indicating that the results from ecological studies have not prompted sufficient changes in policy and management.

Concurrent with advances by ecologists, economists have taken an interest in the problems of invasive species. Much of their interest centres on the trade-offs that must be made to manage invasive species [114]. For example, economists might make the link between a beneficial trade route and the risks from invasions that it causes. They can then ask if and how the trade route can be restricted or modified to reduce risks from invasive species while maximizing overall societal welfare (e.g. [120]).

Although ecologists and economists have often addressed similar concerns (i.e. how to reduce the impacts of invasive species), they have rarely worked together. This stems in part from a lack of understanding between the disciplines. In particular, ecologists may be unwilling to consider ecosystem components/qualities (e.g. biodiversity) in monetary terms. On the economist side, there is often a lack of willingness to work with ecological systems because they are so complex, in many cases are poorly understood even by ecologists, and because the relationships among ecosystem components may not fit into traditional economic models.

As described above, the introduction of non-native species is a consequence of the globalization of trade and travel. The benefits to society from globalization are large, as are the benefits from non-native species used in agriculture, horticulture, forestry, fisheries, landscaping, and the pet trade, etc. Thus, the benefits from policy that prevents and manages invasions must be balanced against the costs of the policy, including the costs of restricting trade. In practical terms, this requires a good understanding of both the ecological and economic dimensions of invasions, and requires that economists and ecologists work together to design appropriate policies.

## General Principles for Policy and Management

It is generally accepted, and has often been shown, that the best way to reduce total impacts from invasive species is to prevent their introduction [113,119,121]. For example, studies have shown that preventing the spread of an invader among lakes [122,123], and the spread of species among countries through trade [120], is economically preferable to managing species after they arrive. If prevention of all introductions is not feasible, efforts to reduce propagule pressure should be the goal as this may effectively prevent a proportion of invasions. Prevention is considered more effective than managing established invaders for at least three reasons. First, it is the only sure way to avoid impacts of invasive species, which as noted above are rarely eradicated. Second, there are more management options available prior to invasion. These include a variety of laws and regulations, quarantine efforts, and the threat of punitive action against individuals or companies that do not comply. Finally, prevention efforts give the opportunity to shift the costs of reducing impacts to the trades that derive greatest benefits from the movement of species [1]. For example, although the pet trade has introduced many harmful invaders around the world, the exporters, importers, and retailers who derive greatest benefits from the trade have not been required to pay the costs. Instead, those costs are generally borne by society as government agencies work to control invaders. Prior to invasion, however, it is possible to impose regulations that require the trades to ensure that they are not transporting potential invaders. Thus, the externality of trade can be internalized.

Post-establishment, the broad range of invasive species and the ecosystems they become established in, and the heterogeneous ways that humans value those ecosystems, means that few generalizations can be given for management and policy. That is, appropriate management and policy for harmful invasive species is highly context dependent. Options include poisoning (e.g. herbicides, pesticides, piscicides), manual removal (e.g. pulling plants), capturing and killing (e.g. trapping, shooting), and release of additional species that may provide biological control of the invader. As the applications of these methods are context dependent, and because different countries within Europe have different standards for the application of these methods, they are not explored further here. The general principle of prevention still applies, however, and efforts to slow or eliminate population spread will often be the most cost-effective ways to reduce total impacts.

# INVASIVE SPECIES POLICY IN EUROPE

There is a large diversity of national approaches to invasive species within the EU (see review in [124]). These range from relatively advanced policies and procedures, such as in the Netherlands where there is a national policy framework for addressing invasive species, to nations that do relatively little to prevent the arrival and spread of invaders. Open internal borders, however, create a weak-link problem in the EU where even nations with the most stringent policies remain at risk because some of their neighbors and trading partners are doing relatively little [1].

At the international level there are several instruments that address the threats from invasive species. These include the Convention on Biological Diversity (CBD, implemented into European legislation by Council Decision 93/626/EEC), which requires ratifying nations to work to prevent the introduction, spread, and export of all types of invasive species (http://www.cbd.int/). There is no history, however, of nations being sanctioned for not following the invasive species guidelines, and there is little evidence that the CBD has led to improved invasive species management at the global scale.

Non-native species that pose direct threats to animals or plants (cultivated or wild) are actively managed at the international level by the World Organization for Animal Health (OIE) and International Plant Protection Convention (IPPC), respectively. While these instruments have proven effective for restricting the spread of some invaders, they have most often been used to address diseases of livestock and pests of crops, with less attention paid to invasive species that cause primarily environmental harm [124]. Within the framework of the IPPC is the European and Mediterranean Plant Protection Convention which develops regional measures to prevent the spread of invasive plants and plant pests for its 50 member nations (i.e. more than just the EU nations). Its focus has also been largely on pests of agriculture.

Although the EU now has policies that address several region-wide environmental issues (e.g. climate change, pollution by chemicals), it has not yet created legislation to address the full range of invasive species and their pathways. Instead, there are several regimes and directives that address some invasive species or some pathways of introduction. We give two examples here. First, Council Regulation 708/2007 sets procedures and standards for aquaculture within the EU with the goal of reducing the risks that non-native fish, or the diseases they carry, will become established and invasive. Second, the Plant Health Regime addresses the introduction and spread of invasive plants and plant pests, but has been used primarily to address pests of agriculture. Plants that pose risks of environmental harm, although they may arrive through the same pathways, are generally not covered.

Although the two examples just given address important aspects of the invasive species problem, the general European policy framework remains spread across many similar and disjunct pieces of legislation. This creates confusion about which species and pathways should be addressed by which instrument, and ultimately means that many invasive species and pathways are not addressed [124]. In response, several programs have been established to identify gaps in the current framework and to suggest solutions. One example is the IMPASSE program that has reviewed the risks of introduction of invasive fish species and fish diseases through aquaculture (http://ec.europa.eu/research/fp6/ssp/impasse_en.htm). However, even if all recommendations from these programs were enacted, the basic European legislative framework would remain scattered, creating large obstacles to a coordinated response to the invasive species problem.

This lack of coordination has been identified as a serious shortcoming by the Commission of the European Communities [14]. In response, the commission set forth three policy options. The first is a business as usual model, in which the EU continues to rely on the current suite of national and EU policies, and the range of international policy instruments, to address invasive species. The second is to maximize use of the existing EU policy framework, possibly with some modifications of legal instruments so that they better address invasive species. This approach appeals because it would not require new legislation. However, it would still be limited in its application because no current policies address all invasive species or pathways. The final option is to create a new legal instrument at the EU level that would apply to all member states. This third option could be complemented by the creation of a European Centre for Invasive Species Management that would provide a coordinating role [125]

Shine et al [124] recently prepared a comprehensive review of existing national, international and European policy, and built upon this to evaluate the three policy options. They strongly recommended the final option - a new legislative instrument - and noted that it is the only option that would provide a basis for coordinated action against all invasive species. Additionally, an economic analysis showed that although this option would be the most expensive, its costs would be much less than the avoided costs of the invasive species that it would prevent from arriving, or that would be managed more efficiently [124]. This recommendation is currently under consideration by the EU, and a final decision about policy approach is expected by 2012.

# CONCLUSIONS

Europe's position as a centre for international trade over many centuries has resulted in it having a large number and diversity of established non-native species. Many of these species are invasive and they affect all European habitats. Impacts include loss of native biodiversity, economic losses, and harm to human health. Over the last several decades the field of invasion biology has grown markedly [19] and has created a growing understanding of the biology of invasive species. One of the emergent insights of invasion biology has been that invasions are an inherently multi-disciplinary issue. More specifically, species are introduced through human economic activities. Without a clear understanding of these activities it is impossible to explain and understand biological invasions. This is clearly illustrated by the importance of propagule pressure as a driver of invasions. Propagule pressure is driven by human activities and is increasingly recognized as a strong factor in many, if not most, invasions.

Where timelines of species non-native species discovery are available, it is clear that the problem of invasions has not been fully addressed by current national or EU policies. Indeed, many European habitats appear to be acquiring new non-native species faster now than at any time in recorded history. Creating and enforcing policies that slow the rate of invasion is difficult for many reasons, including the value derived from many of the pathways that move the most species (e.g. shipping). These policies may be especially difficult in Europe because it is an environmentally heterogeneous region where barriers to the movement of goods have been actively removed to encourage free trade and travel. Despite this difficulty, current deliberations in the EU about developing a comprehensive policy instrument to address invasive species are very promising.

Apart from the policy and management considerations, the spread of non-native species offers many insights into biological and ecological processes as well as into the links between biodiversity, ecosystem functioning, and socio-economic impacts. Many researchers have been attracted to study species invasions because of the complexities, and the possibility to integrate across disciplines in ways that both generate new insight and that can lead to practical policy solutions. By reviewing a broad range of invasive species, and by reviewing both the biological and economic literature, we hope that this manuscript can make a useful contribution toward a more integrative understanding of species invasions in Europe.

# ACKNOWLEDGEMENTS

We thank both the Alexander von Humboldt Foundation and the U. S. National Academy of Sciences for bringing us together at the 16th German-American Frontiers of Science Symposium in Potsdam, Germany, June 2010. We especially thank Henner Hollert for inviting us to write this special review paper, and appreciate helpful comments by Ingo Kowarik and two anonymous reviewers. JMJ acknowledges financial support from the Deutsche Forschungsgemeinschaft (DFG; JE 288/4-1).

# REFERENCES

1. Perrings C, Burgiel S, Lonsdale WM, Mooney H, Williamson M: International cooperation in the solution to trade-related invasive species risks. *Ann NY Acad Sci* 2010, 1195:198-212.
2. Kolar CS, Lodge DM: Progress in invasion biology: predicting invaders. *Trends Ecol Evol* 2001, 16:199-204.
3. Williamson M: *Biological invasions*. London: Chapman & Hall; 1996.
4. Sala OE, Chapin FS, Armesto JJ, Berlow E, Bloomfield J, Dirzo R, Huber-Sanwald E, Huenneke Lf, Jackson RB, Kinzig A, Leemans R, Lodge DM, Mooney HA, Oesterheld M, Poff NL, Sykes MT, Walker BH, Walker M, Wall DH: Global Biodiversity Scenarios for the Year 2100. *Science* 2000, 287:1770-1774.
5. Jeschke JM, Strayer DL: Invasion success of vertebrates in Europe and North America. *P Natl Acad Sci USA* 2005, 102:7198-7202.
6. DAISIE: *Handbook of alien species in Europe*. Dordrecht: Springer; 2009.
7. Vilà M, Basnou C, Pyšek P, Josefsson M, Genovesi P, Gollasch S, Nentwig W, Olenin S, Roques A, Roy D, Hulme PE, DAISIE partners: How well do we understand the impacts of alien species on ecosystem services? A pan-European cross-taxa assessment. *Front Ecol Environ* 2010, 8:135-144.
8. Keller RP, zu Ermgassen PSE, Aldridge D: Vectors and timing of nonindignenous freshwater species establishment in Great Britain. *Conserv Biol* 2009, 23:1526-1534.
9. Hulme PE, Bacher S, Kenis M, Klotz S, Kühn I, Minchin D, Nentwig W, Olenin S, Panov V, Pergl J, Pyšek P, Roques A, Sol D, Solarz W, Vilà M: Grasping at the routes of biological invasions: a framework for integrating pathways into policy. *J Appl Ecol* 2008, 45:403-414.

10. Lambdon PW, Pyšek P, Basnou C, Hejda M, Arianoutsou M, Essl F, Jarosik V, Pergl J, Winter M, Anastasiu P, Andriopoulos P, Bazos I, Brundu G, Celesti-Grapow L, Chassot P, Delipetrou P, Josefsson M, Kark S, Klotz S, Kokkoris Y, Kühn I, Marchante H, Perglova I, Pino J, Vilà M, Zikos A, Roy D, Hulme PE: Alien flora of Europe: species diversity, temporal trends, geographical patterns and research needs. *Preslia* 2008, 80:101-149.

11. Kowarik I: *Biologische Invasionen: Neophyten und Neozoen in Mitteleuropa*. 2nd edition. Stuttgart: Ulmer; 2010.

12. Kowarik I: Time lags in biological invasions with regard to the success and failure of alien species. In *Plant Invasions: General Aspects and Special Problems*. Edited by Pyšek P, Prach K, Rejmánek M, Wade M. Amsterdam: SPB Academic Publishing; 1995:15-38.

13. Essl F, Dullinger S, Rabitsch W, Hulme PE, Huelber K, Jarošík V, Kleinbauer I, Krausmann F, Kühn I, Nentwig W, Vilà M, Genovesi P, Gherardi F, Desprez-Loustau ML, Roques A, Pyšek P: Socio-economic legacy yields an invasion debt. *P Natl Acad Sci USA* 2011, 108:203-207.

14. Commission of the European Communities: Towards a strategy on invasive species.[http:/ / ec.europa.eu/ environment/ nature/ invasivealien/ docs/ 1_EN_ACT_part1_v6.pdf] 2008. Communication from the Commission to the Council, the European Parliament, the European Economic and Social Committee and the Committee of Regions. COM(2008) 789 final

15. Council of the European Union: A mid-term assessment of implementing the EU Biodiversity Action Plan and Towards an EU Strategy on Invasive Alien Species – Council conclusions.[http://register.consilium.europa.eu/pdf/en/09/st11/st11412.en09.pdf] 2009. Information Note 11412/09

16. Lodge DM: Biological invasions: lessons for ecology. *Trends Ecol Evol* 1993, 8:133-137.

17. Sax DF, Stachowicz JJ, Brown JH, Bruno JF, Dawson MN, Gaines SD, Grosberg RK, Hastings A, Holt RD, Mayfield MM, O'Connor MI, Rice WR: Ecological and evolutionary insights from species invasions. *Trends Ecol Evol* 2007, 22:465-471.

18. Richardson DM, (ed): *Fifty Years of Invasion Ecology - The Legacy of Charles Elton*. Chichester (UK): Wiley-Blackwell; 2011.

19. Richardson DM, Pyšek P, Rejmánek M, Barbour MG, Panetta FD, West CJ: Naturalization and invasion of alien plants: concepts and definitions. *Divers Distrib* 2000, 6:93-107.

20. Colautti RI, MacIsaac HJ: A neutral terminology to define 'invasive' species. *Divers Distrib* 2004, 10:135-141.

21. Jeschke JM: Across islands and continents, mammals are more successful invaders than birds. *Divers Distrib* 2008, 14:913-916.

22. Thellung A: La flore adventice de Montpellier. *Mémoires de la Société nationale des Sciences naturelles et mathématiques de Cherbourg*1912, 38:57-728.

23. Thellung A: Pflanzenwanderungen unter dem Einfluss des menschen. *Beibl Bot jahrb Syst* 1915, 53:37-66.

24. Wilson JRU, Dormontt EE, Prentis PJ, Lowe AJ, Richardson DM: Something in the way you move: dispersal pathways affect invasion success. *Trends Ecol Evol* 2009, 24:136-144.

25. Kowarik I, von der Lippe M: Pathways in plant invasions. In *Biological Invasions*. Edited by Nentwig W. Berlin: Springer; 2007:29-47.

26. Holdich DM, Rogers WD, Reynolds JD: Native and alien crayfish in the British Isles. In*Crayfish in Europe as alien species: how to make the best of a bad situation?*. Edited by Gherardi F, Holdich DM. Rotterdam: AA Balkema; 1999:221-235.

27. Benedict MQ, Levine RS, Hawley WA, Lounibos LP: Spread of the tiger: global risk of invasion by the mosquito *Aedes albopictus*. *Vector-Borne Zoonot* 2007, 7:76-85.

28. Keller RP, Drake JM, Drew M, Lodge DM: Linking environmental conditions and ship movements to estimate invasive species transport across the global shipping network. *Divers Distrib* 2011, 17:93-102.

29. Olenin S, Minchin D, Daunys D, Ziako A: Pathways of aquatic invasions in Europe. In*Atlas of Biodiversity Risk*. Edited by Settele J, Penev LD, Georgiev TA, Grabaum R, Grobelnik V, Hammen V, Klotz S, Kotarac M, Kühn I. Sofia (Bulgaria): Pensoft; 2010:138-139.

30. Augustin S, Guichard S, Heitland W, Freise J, Svatoš A, Gilbert M: Monitoring and dispersal of the invading Gracillariidae *Cameraria ohridella*. *J Appl Entomol* 2009, 133:58-66.

31. Genovesi P, Bacher S, Kobelt M, Pascal M, Scalera R: Alien mammals of Europe. In*Handbook of Alien Species in Europe*. Edited by DAISIE. Dordrecht: Springer; 2009:119-128.
32. Kark S, Solarz W, Chiron F, Clergeau P, Shirley S: Alien birds, amphibians and reptiles of Europe. In *Handbook of Alien Species in Europe*. Edited by DAISIE. Dordrecht: Springer; 2009:105-118.
33. Roques A, Rabitsch W, Rasplus JY, Lopez-Vaamonde C, Nentwig W, Kenis M: Alien terrestrial invertebrates of Europe. In *Handbook of Alien Species in Europe*. Edited by DAISIE. Dordrecht: Springer; 2009:63-79.
34. Pyšek P, Lambdon PW, Arianoutsou M, Kühn I, Pino J, Winter M: Alien vascular plants of Europe. In *Handbook of Alien Species in Europe*. Edited by DAISIE. Dordrecht: Springer; 2009:43-61.
35. Bourke PMA: Emergence of potato blight, 1843-46. *Nature* 1964, 203:805-808.
36. Galil BS, Gollasch S, Minchin D, Olenin S: Alien marine biota of Europe. In *Handbook of Alien Species in Europe*. Edited by DAISIE. Dordrecht: Springer; 2009:93-104.
37. Gherardi F, Gollasch S, Minchin D, Olenin S, Panov VE: Alien invertebrates and fish in European inland waters. In *Handbook of Alien Species in Europe*. Edited by DAISIE. Dordrecht: Springer; 2009:81-92.
38. Leuven RSEW, van der Velde G, Baijens I, Snijders J, van der Zwart C, Lenders HJR, bij de Vaate A: The river Rhine: a global highway for dispersal of aquatic invasive species. *Biol Invasions* 2009, 11:1989-2008.
39. Sol D, Vilà M, Kühn I: The comparative analysis of historical alien introductions. *Biol Invasions* 2008, 10:1119-1129.
40. Darwin C: *On the Origin of Species by Means of Natural Selection, or the Preservation of Favoured Races in the Struggle for Life.* London: Murray; 1859.
41. Cadotte MW: Darwin to Elton: early ecology and the problem of invasive species.In *Conceptual Ecology and Invasion Biology: Reciprocal Approaches to Nature*. Edited by Cadotte MW, McMahon SM, Fukami T. Dordrecht: Springer; 2006:15-33.
42. Elton CS: *The ecology of invasions by animals and plants*. London: Methuen; 1958.

43. Levine JM, D'Antonio CM: Elton revisited: A review of evidence linking diversity and invasibility. *Oikos* 1999, 87:15-26.

44. Deutschewitz K, Lausch A, Kühn I, Klotz S: Native and alien plant species richness in relation to spatial heterogeneity on a regional scale in Germany. *Global Ecol Biogeogr* 2003, 12:299-311.

45. Stohlgren TJ, Barnett DT, Kartesz JT: The rich get richer: patterns of plant invasions in the United States. *Front Ecol Environ* 2003, 1:11-14.

46. Fridley JD, Stachowicz JJ, Naeem S, Sax DF, Seabloom EW, Smith MD, Stohlgren TJ, Tilman D, Von Holle B: The invasion paradox: Reconciling pattern and process in species invasions. *Ecology* 2007, 88:3-17.

47. Kühn I, Klotz S: From Ecosystem invasibility to local, regional and global patterns of invasive species. In *Biological Invasions*. Edited by Nentwig W. Heidelberg: Springer; 2007:181-196.

48. Duncan RP, Blackburn TM, Sol D: The ecology of bird introductions. *Annu Rev Ecol Evol S* 2003, 34:71-98.

49. Leprieur F, Beauchard O, Blanchet S, Oberdorff T, Brosse S: Fish invasions in the world's river systems: when natural processes are blurred by human activities. *PLoS Biol* 2008, 6:e28. 2008

50. Blackburn TM, Cassey P, Lockwood JL: The island biogeography of exotic bird species. *Global Ecol Biogeogr* 2008, 17:246-251.

51. Chiron F, Shirley S, Kark S: Human-related processes drive the richness of exotic birds in Europe. *P Roy Soc B-Biol Sci* 2009, 276:47-53.

52. Jeschke JM, Genovesi P: Do biodiversity and human impact influence the introduction or establishment of alien mammals? *Oikos* 2011, 120:57-64.

53. Kühn I, May R, Brandl R, Klotz S: Plant distribution patterns in Germany - Will aliens match natives? *Feddes Repertorium* 2003, 114:559-573.

54. Pino J, Font X, Carbo J, Jove M, Pallares L: Large-scale correlates of alien plant invasion in Catalonia (NE of Spain). *Biol Conserv* 2005, 122:339-350.

55. Pyšek P, Kucera T, Jarosik V: Plant species richness of nature reserves: the interplay of area, climate and habitat in a central European landscape. *Global Ecol Biogeogr* 2002, 11:279-289.

56. Gasso N, Sol D, Pino J, Dana ED, Lloret F, Sanz-Elorza M, Sobrino E, Vilà M: Exploring species attributes and site characteristics to assess plant invasions in Spain. *Divers Distrib* 2009, 15:50-58.
57. Chytrý M, Pyšek P, Wild J, Pino J, Maskell LC, Vilà M: European map of alien plant invasions based on the quantitative assessment across habitats. *Divers Distrib* 2009, 15:98-107. 2009
58. Davis MA, Grime JP, Thompson K: Fluctuating resources in plant communities: a general theory of invasibility. *J Ecol* 2000, 88:528-534.
59. Britton-Simmons KH, Abbott KC: Short- and long-term effects of disturbance and propagule pressure on biological invasion. *J Ecol* 2008, 96:68-77.
60. di Castri F: History of biological invasions with special emphasis on the Old World.In *Biological Invasions: A Global Perspective*. Edited by Drake JA, Mooney HA, di Castri F, Groves RH, Kruger FJ, Rejmánek M, Williamson M. Chichester UK: John Wiley and Sons; 1989:1-30.
61. Becker T, Dietz H, Billeter R, Buschmann H, Edwards PJ: Altitudinal distribution of alien plant species in the Swiss Alps. *Perspect Plant Ecol* 2005, 7:173-183.
62. Pyšek P, Jarošík V, Hulme PE, Kühn I, Wild J, Arianoutsou M, Bacher S, Chiron F, Didžiulis V, Essl F, Genovesi P, Gherardi F, Hejda M, Kark S, Lambdon PW, Desprez-Loustau ML, Nentwig W, Pergl J, Poboljšaj K, Rabitsch W, Roques A, Roy DB, Shirley S, Solarz W, Vilà M, Winter M: Disentangling the role of environmental and human pressures on biological invasions across Europe. *P Natl Acad Sci USA* 2010, 107:12157-12162.
63. Galil BS: A sea under siege - alien species in the Mediterranean. *Biol Invasions* 2000, 2:177-186.
64. Leppäkoski E, Olenin S, Gollasch S: The Baltic Sea: a field laboratory for invasion biology. In *Invasive Aquatic Species of Europe: Distribution, Impacts and Management*. Edited by Leppäkoski E, Gollasch S, Olenin S. Dordrecht: Kluwer; 2002:253-259.
65. Panov VE, Bychenkov DE, Berezina NA, Maximov AA: Alien species introductions in the eastern Gulf of Finland: current state and possible management options. *Proc Est Acad Sci Biol Ecol* 2003, 52:254-267.

66. Panov V, Dgebuadze Y, Shiganova T, Filippov A, Minchin D: A risk assessment of biological invasions: inland waterways of Europe - the northern invasion corridor case study. In *Biological Invaders in Inland Waters: Profiles, Distribution, and Threats*. Edited by Gherardi F. Dordrecht: Springer; 2007:639-656.
67. Wolff WJ, Reise K: Oyster imports as a vector for the introduction of alien species into northern and western European coastal waters. In *Invasive Aquatic Species of Europe: Distribution, Impacts and Management*. Edited by Leppäkoski E, Gollasch S, Olenin S. Dordrecht: Kluwer; 2002:193-205.
68. Baker HG: The evolution of weeds. *Annu Rev Ecol Syst* 1974, 5:1-24.
69. Jeschke JM, Strayer DL: Determinants of vertebrate invasion success in Europe and North America. *Global Change Biol* 2006, 12:1608-1619.
70. Pyšek P, Richardson DM: Traits Associated with Invasiveness in Alien Plants: Where Do we Stand? In *Biological Invasions*. Edited by Nentwig W. Dordrecht: Springer; 2007:97-125.
71. Genovesi P, Shine C: *European strategy on invasive alien species*. Strasbourg: Council of Europe Publishing; 2004.
72. Weber E, Köhler B, Gelpke G, Perrenoud A, Gigon A: Schlüssel zur Einteilung von Neophyten in der Schweiz in die Schwarze Liste oder die Watch-Liste. *Bot Helv* 2005, 115:169-194.
73. Gederaas L, Salvesen I, Viken Å: *Norwegian Black List 2007: Ecological risk analysis of alien species*. Trondheim: Artsdatabanken; 2007.
74. Parrott D, Roy S, Baker R, Cannon R, Eyre D, Hill M, Wagner M, Preston C, Roy H, Beckmann B, Copp GH, Edmons N, Ellis J, Laing I, Britton R, Gozlan RE, Mumford J: *Horizon scanning for new invasive non-native animal spe-cies in England*. Sheffield: Natural England; 2009.
75. Nehring S, Essl F, Klingenstein F, Nowack C, Rabitsch W, Stöhr O, Wiesner C, Wolter C:*Schwarze Liste invasiver Arten: Kriteriensystem und Schwarze Listen invasiver Fische für Deutschland und für Österreich*. Bonn (Germany): Bundesamt für Naturschutz; 2010.

76. Keller RP, Drake JM: Trait based risk assessment for invasive species. In*Bioeconomics of Invasive Species: Integrating Ecology, Economics, Policy and Management*. Edited by Keller RP, Lodge DM, Lewis MA, Shogren JF. New York: Oxford University Press; 2009:44-62.
77. Hayes KR, Barry SC: Are there any consistent predictors of invasion success? *Biol Invasions* 2008, 10:483-506.
78. Lockwood JL, Cassey P, Blackburn T: The role of propagule pressure in explaining species invasions. *Trends Ecol Evol* 2005, 20:223-228.
79. Blackburn TM, Duncan RP: Establishment patterns of exotic birds are constrained by non-random patterns in introduction. *J Biogeogr* 2001, 28:927-939.
80. Sol D, Duncan RP, Blackburn TM, Cassey P, Lefebvre L: Big brains, enhanced cognition, and response of birds to novel environments. *P Natl Acad Sci USA* 2005, 102:5460-5465.
81. Cassey P, Blackburn T, Russell GJ, Jones KE, Lockwood JL: Influences on the transport and establishment of exotic bird species: an analysis of the parrots (Psittaciformes) of the world. *Global Change Biol* 2004, 10:417-426.
82. Blackburn TM, Cassey P, Lockwood JL: The role of species traits in the establishment success of exotic birds. *Global Change Biol* 2009, 15:2852-2860.
83. Hanspach J, Kühn I, Pyšek P, Boos E, Klotz S: Correlates of naturalization and occupancy of introduced ornamentals in Germany. *Perspect Plant Ecol* 2008, 10:241-250.
84. Williamson M, Dehnen-Schmutz K, Kühn I, Hill M, Klotz S, Milbau A, Stout J, Pyšek P: The distribution of range sizes of native and alien plants in four European countries and the effects of residence time. *Divers Distrib* 2009, 15:158-166.
85. Pyšek P, Jarošik V, Pergl J, Randall R, Chytrý M, Kühn I, Tichy L, Danihelka J, Jun JC, Sádlo J: The global invasion success of Central European plants is related to distribution characteristics in their native range and species traits. *Divers Distrib* 2009, 15:891-903. 2009
86. van Kleunen M, Weber E, Fischer M: A meta-analysis of trait differences between invasive and non-invasive plant species. *Ecol Lett* 2010, 13:235-245.

87. Dawson W, Fischer M, van Kleunen M: The maximum relative growth rate of common UK plant species is positively associated with their global invasiveness. *Global Ecol Biogeogr* 2011, 20:299-306.

88. Küster EC, Kühn I, Bruelheide H, Klotz S: Trait interactions help explain plant invasion success in the German flora. *J Ecol* 2008, 96:860-868.

89. Thompson K, Davis MA: Why research on traits of invasive plants tells us very little. *Trends Ecol Evol* 2011, 26:155-156.

90. Paavola M, Olenin S, Leppäkoski E: Are invasive species most successful in habitats of low native species richness across European brackish water seas? *Estuar Coast Shelf S* 2005, 64:738-750.

91. Strayer D, Downing JA, King TL, Layer JB, Newton TJ, Nichols SJ: Changing perspectives on pearly mussels, North America's most imperiled animals. *BioScience* 2004, 54:429-439.

92. Geist J: Strategies for the conservation of endangered freshwater pearl mussels (*Margaritifera margaritifera* L.): a synthesis of Conservation Genetics and Ecology. *Hydrobiologia* 2010, 644:69-88.

93. Keller RP, Drake JM, Lodge DM: *Fecundity as a basis for risk assessment of nonindigenous freshwater molluscs. Conserv Biol.* 2007, 21:191-200.

94. Karatayev AY, Burlakova LE, Padilla DK, Matitsky SE, Olenin S: Invaders are not a random selection of species. *Biol Invasions* 2009, 11:2009-2019.

95. Strayer DL, Eviner VT, Jeschke JM, Pace ML: Understanding the long-term effects of species invasions. *Trends Ecol Evol* 2006, 21:645-651.

96. Bisby FA, Roskov YR, Orrell TM, Nicolson D, Paglinawan LE, Bailly N, Kirk PM, Bourgoin T, Baillargeon G: Species 2000 & ITIS Catalogue of Life: 2010 Annual Checklist.[http://www.catalogueoflife.org/annual-checklist/2010] Accessed 30 July 2010

97. Pyšek P, Richardson DM, Rejmánek M, Webster BL, Williamson M, Kirschner J: Alien plants in checklists and floras: towards

better communication between taxonomists and ecologists. *Taxon* 2004, 53:131-143.

98. Gherardi F, Bertolino S, Bodon M, Casellato S, Cianfanelli S, Ferraguti M, Lori E, Mura G, Nocita A, Riccardi N, Rossetti G, Rota E, Scalera R, Zerunian S, Tricarico E: Animal xenodiversity in Italian inland waters: distribution, modes of arrival, and pathways. *Biol Invasions* 2008, 10:435-454.
99. Kettunen M, Genovesi P, Gollasch S, Pagad S, Starfinger U, ten Brink P, Shine C: *Technical support to EU strategy on invasive alien species (IAS) - Assessment of the impacts of IAS in Europe and the EU.* Brussels: Institute for European Environmental Policy; 2009.
100. Kumschick S, Nentwig W: Some alien birds have as severe an impact as the most effectual alien mammals in Europe. *Biol Conserv* 2010, 143:2757-2762.
101. Nentwig W, Kühnel E, Bacher S: A generic impact-scoring system applied to alien mammals in Europe. *Conserv Biol* 2010, 24:302-311.
102. Vilà M, Weber E, D'Antonio CM: Conservation implications of invasion by plant hybridization. *Biol Invasions* 2000, 2:207-217.
103. Pyšek P, Cock MJW, Nentwig W, Ravn HP, (Eds): *Ecology and management of giant hogweed (Heracleum mantegazzianum).* Wallingford (UK): CABI; 2007.
104. Reinhardt F, Herle M, Bastiansen F, Streit B: *Ökonomische Folgen der Ausbreitung von gebietsfremden Organismen in Deutschland.* Berlin: Umweltbundesamt; 2003.
105. Andreu J, Vilà M, Hulme PE: An assessment of stakeholder perceptions and management of noxious alien plants in Spain. *Environ Manage* 2009, 43:1244-1255.
106. Dudgeon D, Arthington AH, Gessner MO, Kawabata ZI, Knowler DJ, Lévêque C, Naiman RJ, Prieur-Richard AH, Soto D, Stiassny MLJ, Sullivan CA: Freshwater biodiversity: importance, threats, status and conservation challenges. *Biol Rev* 2006, 81:163-182.
107. Geist J: Integrative freshwater ecology and biodiversity conservation. *Ecol Indic* 2011, in press.

108. Souty-Grosset C, Holdich DM, Noel PY, Reynolds JD, Haffner P, (Eds): *Atlas of crayfish in Europe*. Paris: Museum National d'Histoire Naturelle; 2006. Patrimonines Naturels 64

109. Söderbäck B: Replacement of the native crayfish *Astacus astacus* by the introduced species *Pacifastacus leniusculus* in a Swedish lake: possible causes and mechanisms. *Freshwater Biol* 1995, 33:291-304.

110. Söderhäll K, Cerenius L: The crayfish plague fungus: history and recent advances. *Freshwater Crayfish* 1999, 12:11-35.

111. Rahel FJ: Homogenization of freshwater faunas. *Ann Rev Ecol Syst* 2002, 33:291-315.

112. Longepierre S, Robert A, Levi F, Francour P: How an invasive alga species (*Caulerpa taxifolia*) induces changes in foraging strategies of the benthivorous fish *Mullus surmuletus* in coastal Mediterranean ecosystems. *Biodivers Conserv* 2005, 14:365-376.

113. Lodge DM, Williams S, MacIsaac H, Hayes K, Leung B, Loope L, Reichard S, Mack RN, Moyle PB, Smith M, Andow DA, Carlton JT, McMichael A: Biological invasions: recommendations for policy and management. *Ecol Appl* 2006, 16:2035-2054.

114. Keller RP, Lodge DM, Lewis M, Shogren J: *Bioeconomics of Invasive Species: Integrating Ecology, Economics, Policy and Management*. New York: Oxford University Press; 2009.

115. Molnar JL, Gamboa RL, Revenga C, Spalding MD: Assessing the global threat of invasive species to marine biodiversity. *Front Ecol Environ* 2008, 6:485-492.

116. Ricciardi A: Patterns of invasion in the Laurentian Great Lakes in relation to changes in vector activity. *Divers Distrib* 2006, 12:425-433.

117. Carlton J: Predictions of the arrival of the zebra mussel in North America. *Dreissena Polymorpha Information Review* 1991, 2:1.

118. Finnoff D, Shogren JF, Leung B, Lodge D: Take a risk: preferring prevention over control of biological invaders. *Ecol Econ* 2007, 62:216-222.

119. IUCN: IUCN Guidelines for the Prevention of Biodiversity Loss Caused by Alien Invasive Species. [http://iucn.org] Gland

(Switzerland): International Union for the Conservation of Nature; 2000.

120. Keller RP, Lodge DM, Finnoff DC: Risk assessment for invasive species produces net bioeconomic benefits. *P Natl Acad Sci USA* 2007, 104:203-207.

121. Miller C, Kettunen M, Shine C: *Scope options for EU action on invasive alien species (IAS)*.Final report for the European Commission. Institute for European Environmental Policy (IEEP), Brussels, Belgium; 2006.

122. Leung B, Lodge DM, Finnoff D, Shogren JF, Lewis MA, Lamberti G: An ounce of prevention or a pound of cure: bioeconomic risk analysis of invasive species. *P Roy Soc B-Biol Sci* 2002, 269:2407-2413.

123. Keller RP, Frang K, Lodge DM: Preventing the spread of invasive species: intervention guided by ecological predictions leads to economic benefits. *Conserv Biol* 2008, 22:80-88.

124. Shine C, Kettunen M, Genovesi P, Essl F, Gollasch S, Rabitsch W, Scalera R, Starfinger U, ten Brink P: *Assessment to support continued development of the EU Strategy to combat invasive alien species. Final Report for the European Commission*. Institute for European Environmental Policy (IEEP), Brussels, Belgium; 2010.

125. Hulme PE, Pyšek P, Nentwig W, Vilà M: Will threat of biological invasions unite the European Union? *Science* 2009, 324:40-41.

126. Essl F, Lambdon P: Alien bryophytes and lichens of Europe. In *Handbook of Alien Species in Europe*. Edited by DAISIE. Dordrecht: Springer; 2009:29-41.

127. Gozlan E, Andreou D, Asaeda T, Beyer K, Bouhadad R, Burnard D, Caiola N: Pan-continental invasion of *Pseudorasbora parva*: towards a better understanding of freshwater fish invasions. *Fish Fish* 2010, 11:315-340.

128. Bernauer D, Jansen W: Recent invasions of alien macroinvertebrates and loss of native species in the upper Rhine river, Germany. *Aquat Invasions* 2006, 1:55-71.

129. Van der Velde G, Rajagopal S, Kelleher B: Ecological impact of crustacean invaders: general considerations and examples from the Rhine river.*Crustacean Iss* 2000, 12:3-33.

130. Pöckl M: Success of the invasive Ponto-Caspian amphipod *Dikerogammarus villosus* by life history traits and reproductive capacity.*Biol Invasions* 2009, 11:2021-2041.

131. Bij de Vaate A, Jazdzewski K: Geographical patterns in range extension of Ponto-Caspian macroinvertebrate species in Europe. *Can J Fish Aquat Sci* 2002, 59:1159-1174.

132. Van derVelde G, Platvoet D: Quagga mussels *Dreissena rostiformis* bugensis (Andrusov, 1897) in the Main River (Germany).*Aquat Invasions* 2007, 2:261-264.

133. Figuerola J, Green AJ: Dispersal of aquatic organisms by waterbirds: a review of past research and priorities for future studies.*Freshwater Biol* 2002, 47:483-494.

134. Hussner A, Van de Weyer K, Gross EM, Hilt S: *Comments on increasing number and abundance of non-indigenous aquatic macrophyte species in Germany*. 2010, 50:519-526.

Chapter 3

# Envisioning South-South Relations in the Fields of Environmental Change and Migration in the Pacific Islands - Past, Present and Futures

Eberhard Weber

School of Geography, Earth Science and Environment, the University of the South Pacific, Laucala Campus, Suva, Fiji

## ABSTRACT

Climate change poses severe threats to developing countries. Scientists predict entire states (e.g. Kiribati, Marshall Islands, Tuvalu, and Maldives) will become inhabitable. People living in these states have to resettle to other countries. Media and politicians warn that climate change will trigger migration flows in dimensions unknown to date. It is feared that millions from developing countries overwhelm developed societies and increase pressures on anyway ailing social support systems destabilizing societies and becoming a potential source of conflict.

Inhabitants of Pacific Islands[a] have been mobile since the islands were first settled not longer than 3,500 years ago. Since then people moved around, expanded their reach, and traded with neighbouring tribes (and later countries). With the event of European powers in the 15th century independent mobility became restricted after the beginning of the 19th century. From the second half of the 19th century movements of people predominately served economic interests of colonial powers, in particular a huge colonial appetite for labour. After independence emigration from Pacific Island countries continued to serve economic interest of metropolitan countries at the rim of the Pacific Ocean, which are able to direct migration flows according to their economic requirements.

[a]The term Pacific Islands countries in the context of this paper refers to 15 Pacific ACP states; 14 of them are organized in the Pacific Island Forum Secretariat (PIFS) plus Timor Leste, an island state that is geographically situated in Asia.

If climate change resettlements become necessary in big numbers then Pacific Islanders do not want to become climate change refugees. To include environmental reasons in refugee conventions is not what Pacific Islanders want. They want to migrate in dignity, if it becomes unavoidable to leave their homes. There are good reasons to solve the challenges within Pacific Island societies and do not depend too much on metropolitan neighbours at the rim of the Pacific such as Australia, New Zealand and the USA. To rise to the challenge requires enhanced Pan-Pacific Island solidarity and South-South cooperation. This then would result in a reduction of dependencies. For metropolitan powers still much can be done in supporting capacity building in Pacific Island countries and helping the economies to proposer so that climate change migrants easier can be absorbed by expanding labour markets in Pacific Island countries.

# INTRODUCTION

## Setting the Challenges

Climate Change has become one of the most pressing challenges of the 21st century. For many years it has been seen from a perspective of mitigation: what can be done to prevent climate change from happening? Today's major concerns are: how can societies adjust, adapt to or at least cope with the impacts arising from climate change? In this discourse the question is, if moving away is a suitable adaptation strategy, or an acknowledgement that adaptation has failed.

In the past mobility played an outstanding role in the lives of Pacific Islanders. After the first had arrived some 3,500 years ago they then moved around freely. Later, colonial powers in the request for increasing labour shifted hundreds of thousands of people within the Pacific Island region. When supply of labourers within Pacific Islands faded they brought migrant workers from outside. Once Pacific Island countries became independent the mobility of its people did not abate; many more moved to countries at the rim of the Pacific.

Today the threat of climate change, sea-level rise and the risk of becoming inhabitable are real issues for many Pacific Islands. New motivations to go elsewhere appear on the horizon; yet there are little ideas how resettlement can be managed, if and once required. The paper thus is a beginning to envision how to avoid that climate change migrants become refugees. Migrant receiving countries do not wish to be flooded by huge waves of refugees putting strain on their social, economic and political systems. At the same time no migrant prefers to be a refugee; they all wish to be welcome wherever they have to turn to when their time at home runs out. All wish to be able to support themselves and their families through the skills they have acquired. No one wishes to live in uncertainty, without proper means of livelihoods and threatened to be sent away or punished just because it wasn't possible to stay in the countries they came from.

## The Pacific Island Region

The Pacific Ocean covers more than 165 million square kilometres extending to about a third of the surface of the Earth. With this it is the biggest geographical entity on Planet Earth. The Pacific Islands consists of 20,000 – 30,000 islands located south of the Tropic of Cancer (Ridgell [2006]). As a rough indication the region is sub-divided into Micronesia, Melanesia and Polynesia, and Timor Leste, which lies further west to the actual Pacific islands. The islands of the first sub-division are mainly situated north of the equator and those of the second and third category as well as Timor Leste are situated in the southern hemisphere (Figure 1).

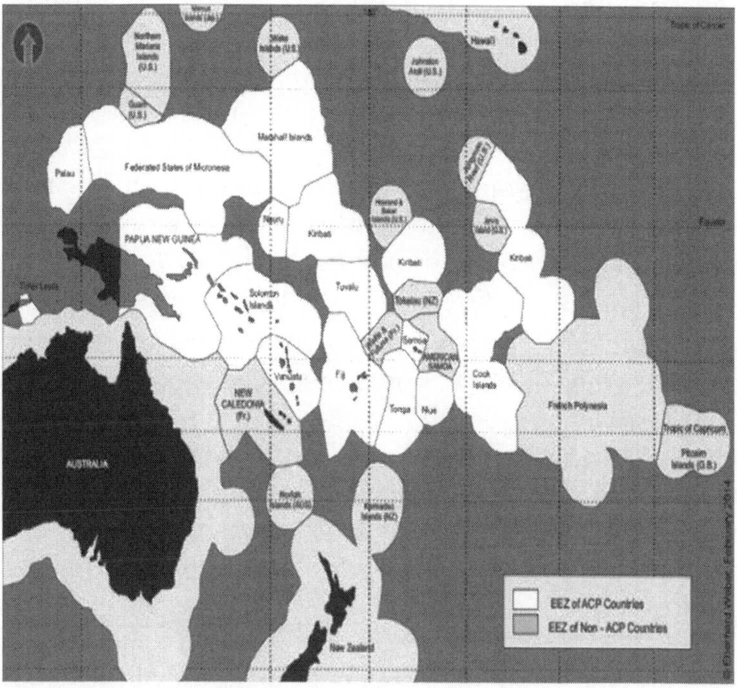


**Figure 1**: Pacific island countries and territories and their exclusive economic zones.

All except one (Tonga[b]) had been colonies of foreign powers. Compared to the Americas and Asia colonialism started rather late and for most Pacific Islands the colonial period didn't last even 100 years.

The first to become independent was Samoa in 1962 and the last Timor Leste where a civil war after a referendum in 1999 delayed formal independence from Indonesia until 2002.

[b]Throughout the paper the modern names of island countries and territories are used.

Today most islands in the region belong to 15 independent states. Some islands, namely those belonging to American Samoa (USA), French Polynesia (France), Tokelau (New Zealand), Wallis & Futuna (France), Guam (USA), Northern Mariana Islands (USA), and New Caledonia (France) are governed by a metropolitan power. Two Pacific ACP countries (Cook Islands and Niue) are in a free association with New Zealand. This includes New Zealand citizenship. Three Pacific ACP countries (Federated States of Micronesia, the Marshall Islands, and Palau) form a Compact of Free Association (COFA) with the USA. The citizens of these countries are free to live and work in the United States (Hills [1984]).

Pacific Island countries and territories (incl. Timor Leste) comprise a total land area of a bit more than 565,000 $km^2$ (Table 1). With their combined land areas they are roughly as big as Kenya. Papua New Guinea, the world's second biggest island next to Greenland, alone has more than 80 percent leaving the remaining 14 states and seven territories with a bit more than 100,000 $km^2$, about the size of South Korea.

**Table 1:** Population, area and island topography of Pacific Islands countries and territories

| Country | Sub-Region | Population (2013) | Population (around 1960) | Land Area (km²)[1] | Exclusive Economic Zone (km²)[1] | Population Density (per/km²) | Atolls/ Coral Islands | Raised Islands | Volcanic Islands | Total |
|---|---|---|---|---|---|---|---|---|---|---|
| Cook Islands | Polynesia | 15,200 | 18,378[f] | 240 | 1,960,135 | 63 | 7 | 0 | 8 | 15 |
| Niue | | 1,500 | 4,864[f] | 259 | 316,629 | 6 | 0 | 1 | 0 | 1 |
| Samoa | | 187,400 | 114,427[f] | 2,934 | 131,812 | 64 | 0 | 0 | 2 + 8 islets | 2 |
| Tonga | | 103,300 | 56,383[a] | 688 | 664,853 | 150 | a few | >100 | a few | ~170 |
| Tuvalu | | 10,900 | 5,444[h] | 26 | 751,797 | 419 | 9 | 0 | 0 | 9 |
| Fed. States of Micronesia | Miconesia | 103,000 | 39,284[c] | 700 | 2,992,597 | 147 | ~600 | 0 | >10 | 607 |
| Kiribati | | 180,800 | 43,336[h] | 726 | 3,437,345 | 149 | 32 | 1 | 0 | 33 |
| Marshall Islands | | 54,200 | 13,928[c] | 181 | 1,992,232 | 299 | 34 | 0 | 0 | 34 |
| Nauru | | 10,500 | 4,613[f] | 21 | 308,502 | 500 | 0 | 1 | 0 | 1 |
| Palau | | 17,800 | 9,344[c] | 487 | 604,289 | 37 | <300 | >10 | >5 | 340 |
| Fiji Islands | Melanesia | 859,200 | 345,737[a] | 18,272 | 1,281,122 | 47 | 2 | a few | <100 | ~320 |
| Papua New Guinea | | 7,398,500 | 2,184,986[i] | 462,000 | 2,396,214 | 16 | a few | a few | >600 | >600 |
| Solomon Islands | | 610,800 | 124,076[d] | 28,000 | 1,597,492 | 22 | a few | 0 | >900 | >990 |
| Vanuatu | | 264,700 | 78,088[k] | 12,190 | 827,891 | 22 | 0 | | 82 | 82 |
| Timor Leste | | 1,066,409 | n.k. | 15,007 | 77,256 | 71 | 0 | 0 | 2 | 2 |

| | | | | | | | | | | |
|---|---|---|---|---|---|---|---|---|---|---|
| sub-total | | 10,812,209 | n.k. | 541,731 | 16,347,569 | | | | | |
| Territory | | | | | | | | | | |
| American Samoa | Poly | 56,500 | 20,051[e] | 199 | 404,391 | 284 | 2 | 0 | 5 | 7 |
| French Polynesia | Poly | 261,400 | 84,551[g] | 3,660 | 4,767,242 | 71 | ~80 | few | ~40 | 130 |
| Tokelau | Poly | 1,200 | 1,870[f] | 12 | 319,031 | 100 | 3 | 0 | 0 | 3 |
| Wallis & Futuma | Poly | 12,200 | 8,546[m] | 143 | 258,269 | 85 | 0 | 0 | 2 | 2 |
| Guam | Micro | 174,900 | 67,044[e] | 549 | 221,504 | 318 | 0 | 0 | 1 | 1 |
| Northern Mariana Islands | Micro | 55,700 | 8,290[c] | 457 | 749,268 | 121 | 0 | 0 | 15 | 15 |
| New Caledonia | Mela | 259,000 | 86,519[c] | 18,600 | 1,422,543 | 14 | 0 | 0 | 7 | 7 |
| sub-total | | 820,900 | n.k. | 23,620 | 8,142,248 | n.k | n.k | n.k | n.k | n.k |
| overall | | 11,633,109 | n.k. | 565,351 | 24,489,817 | n.k | n.k | n.k | n.k | n.k |

[a]1956 [b]1957 [c]1958 [d]1959 [e]1960 [f]1961 [g]1962 [h]1963 [i]1966 [k]1967 [l]1968 [m]1969

Source: for Population Data: SPC Population Data 2011 and Time Series from 1900 for Island size and island type: Pacific Islands Yearbook, 1989.

[1]Land Area and EEZ for Pacific Island Countries according to http://www.forumsec.org/pages.cfm/about-us/our-partners/member-countries/ EEZ for Pacific Island Territories: http://www.seaaroundus.org/eez/ figures for Timor Leste from hp://dnemofgov.tl. Weber *Bandung: Journal of the Global South* 2015 **2**:6, doi: 10.1186/s40728-014-0009-z

On the other hand the water body around the islands is tremendous: the combined Exclusive Economic Zone (EEZ) of all 15 Pacific Island countries and territories is almost 25 million km$^2$. With this they cover an area considerably bigger than Russia, the world's largest country and almost three times the land area of China. Distances within the Pacific Island region are huge and transportation costs, both for people as well as goods, are expensive, especially as a result of rising costs for fossil fuels.

Pacific Islanders have been amongst the most mobile peoples over centuries. Take the tiny country of Niue, which today has a population of not even 1,500 and where another 6,000 live mainly in New Zealand. Or the people of the Cook Islands: some 15,000 of them live at home, whilst more than the same number live around the Pacific Rim, mainly in New Zealand (Campbell and Bedford [2014]). Also other countries, where no dual citizenship with one of the metropolitan countries in their vicinity exists, have meaningful sections of their population outside their territories. Samoa, with a population at home of around 190,000, has another 120,000 people living overseas, mainly in New Zealand (World Bank [2011]). Similarly, Tonga with a bit more than 100,000 people back home and roughly 50,000 outside (World Bank [2011]). Even Fiji, a country that until recently was not perceived as an important source of migrants, has experienced emigration of around 120,000 citizens since the coups of 1987. In total more than 180,000 Fijians live outside their country (World Bank [2011]). This is a huge number, considering that the country in 2014 had a population of less than 900,000.

## The Four Stages of Mobility in Pacific Island Countries

The Pacific Islands were first settled by humans from East Asia, who came to the Pacific Islands in several phases. The first arrived some

40,000 years ago in what today is called New Guinea. They spread to the Bismarck Archipelago some 7,000 years later and arrived in what are the Solomon Islands today some 29,000 years ago. These islands belong to Near Oceania (Matisoo-Smith and Robins [2004]). More than 20,000 years later Remote Oceania was settled from West to East starting from the Reef / Santa Cruz group in the Solomon Islands through Vanuatu, Fiji, Samoa and Tonga. After a pause of 500 –1,000 years Polynesians then settled the rest of the Polynesian Triangle spreading northwards to Hawaii, and southwards to New Zealand (Matisoo-Smith and Robins [2004]: 9167).

With the arrival of Europeans in the Pacific Ocean a second stage started. In the early 16$^{th}$ century the Pacific Oceans was little more than a seemingly endless body of water between the Spanish colonies in Asia and the Americas. During this time many islands were added to European knowledge, their locations drawn into the maps of the Pacific Ocean. It took until the start of the 19$^{th}$ century before the islands first slowly and then since the mid-19$^{th}$ century more quickly were economically integrated into a colonial world system.

With the event of colonial rule movements within the Pacific suddenly became subordinated to external interests. Munro ([1990]: xxxix) estimates that the 'head-count', which is conservative and limited in its own way, enumerated almost one million Pacific Islanders and a further half million Asians who have been involved in the Pacific labour trade between the 1860s and 1939 (Figure 2).

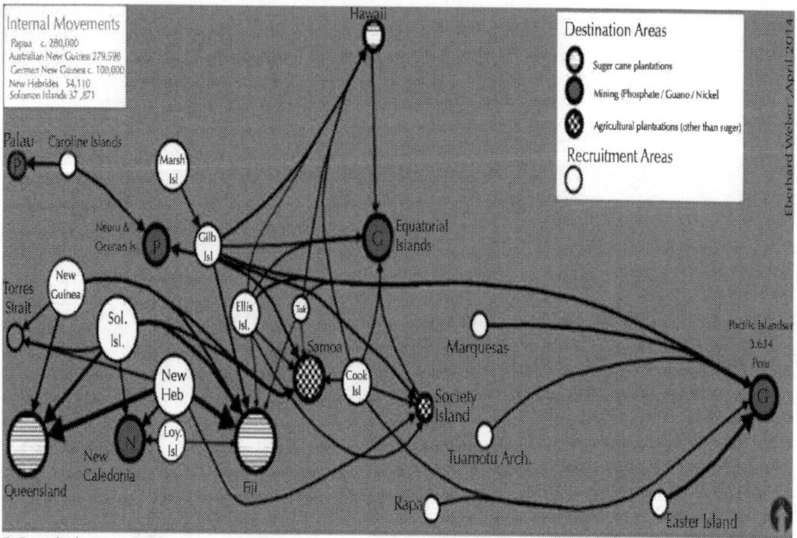


| To Queensland | To Fiji | TO Samoa | To Nauru & Ocean Island | To New Caledonia | To Hawai'i |
|---|---|---|---|---|---|
| **Intra Pacific Island labour Flows in the 19th and early 20th century (incl. Queensland and Hawai'i)** | | | | | |
| New Hebrides 39,931 | New Hebrides 14,198 | New Guinea / Solomon Islands c.10,000 | Gilbert Islands c.1,000 | New Hebrides c.10 -13,000 | Pacific Islanders 2,444 |
| Solomon Islands 18,217 | Solomon Islands 8,228 | Micronesians c. 2,500 | Caroline Islands c. a few hundreds (1) | Solomon Islands c. 1,000 | |
| New Guinea (Ger, Aus) 2,809 | Gilbert Islands 2,398 | Gilbert Islands 210 | Ellis Islands c.300 | | |
| Loyalty Islands 1,123 | New Guinea 1,618 | | | | |
| other Pacific Islands c. 400 | other Pacific Islands 585 | | | | |
| **Labour Flows to the Pacific Islands from outside in the 19yh and early 20th century (incl. Queensland and Hawai'i)** | | | | | |
| | India 60,965 | China 6,984 | China c. 1,200 | French convicts c. 22.000 | Japan 186,343 |
| | Japan 305 | | | Indonesia c.33,000 | Philippines 102,069 |
| | | | | India 376 | China 61,794 |
| | | | | other Asian c.14,000 | Portugal 16,318 |
| | | | | | other European 13,553 |
| | | | | | other Asian 8,283 |
| | | | | | Puerto Rico 5,203 |

Sources: data from Moore et al. (1990) unless indicated otherwise; (1) Firth (1978).

**Figure 2:** Colonial labour movements in the Pacific Island region [Moore et al.[1990]].

Some exotic products like sandalwood (Shineberg [1967]), sea-cucumber (Beche-de-mer; Ward [1972]) caught the attention of European traders, castaways and adventurers and became important commodities for European trade with China. Once these exotic resources had been depleted colonial plantation agricultural for copra, cotton and sugar cane, to name the most important, provided an economic base for an increasing number of European settlers arriving in the bigger islands. To operate these plantations a big number of cheap indentured (contract) labourers was required.

In a few islands colonial plantations were competing with mining activities, for investment, but more importantly for labour. Very early nickel mining had been taken up in New Caledonia (1860s). Phosphate mining in Banaba (Ocean Island, today Kiribati), Nauru, Makatea (today French Polynesia), and Angaur (today Palau) starting at the early 20th century. This brought many Pacific Islanders, but also labourers from Asia to these phosphate islands (Firth [1978]).

Labour mobility in the mid-19th century led to a distinct demographic reconfiguration of many Pacific Island societies. Today they have meaningful sections from other Pacific societies, and meaningful Asian population segments.

In the second half of the nineteenth century some 60,000 Pacific Islanders were taken to Australia by black-birding ships to work in Queensland's sugar plantations (Flanagan et al. [2003]). Around 10,000 were still living in Australia when the country achieved independence in 1901. Most of them were repatriated between 1906 and 1908 as a result of the White Australia Policy. Some 7,500 Pacific Islanders (called *Kanakas*) working on plantations in Queensland were deported, and entry into Australia was prohibited for Pacific Islanders after 1904 (Jupp [2007]).

The third stage happened parallel to the second, but is different in a number of aspects: colonial powers resettled populations of entire islands within their colonial empire. Such resettlements happened to the people of Banaba (today Kiribati), who were brought to Rabi Islands (today Fiji) in 1945, Gilbert Islanders (today Kiribati) who were brought in 1938 to the Phoenix Islands, and then in the 1950s to the Western Province of today's Solomon Islands (Campbell [2014]). This form of migration seems to resemble best what people often think of when they speak of resettlement of islands as a result of climate change: to evacuate the population of entire islands and bring them – as a group - to their new home.

The fourth stage started towards the end of colonial rule and continues into the years of independence. When in the 1950s and 1960s in Germany *Gastarbeiter* (guest workers) and migrants from Mexico into the USA supported the economic miracle of the Post War era also in Australia and New Zealand guest workers were needed in rapidly expanding economies (Ward [1989]). A substantive number of skilled migrants came from Pacific Island countries, where highly skilled professionals were rather scarce. It meant a huge loss for sending countries when many of those few professionals left for greener pastures. At this time all Pacific Island countries were still colonies of European powers, Australia, New Zealand and the USA. Later when most of the countries had achieved independence recruitment of Pacific Islanders continued. Even attempts were made to liberalize immigration policies. New Zealand e.g. granted free entry to Pacific Islanders in 1986 but the program was discontinued the very same year as too many immigrants arrived (Goss and Lindquist [2000]: 393).

## Analysis of Labour Markets in Pacific Island ACP Countries

Labour migration from and to Pacific Island countries reflect on their labour markets requirements, their need to recruit skilled labour from outside, and their ability to provide labour to other Pacific Island countries.

Economies of Pacific Island countries are very little diversified. Generally the formal sector is tiny, and it often consists of a dominating public sector (Table 2). Most wholesale businesses are part of the formal sector, while retail businesses are often semi-formal or informal (Voigt-Graf [2008]). They might require licenses to operate, but other requirements for the formal sector like social security provisions, formal entry requirements etc. are either not necessary or not implemented. Only a few Pacific Island countries have a meaningful modern manufacturing sector concentrating mainly on food processing (incl. alcoholic and non-alcoholic beverages), furniture, textiles and construction materials. The production of consumer electronics (television, computer, fridges, washing machine, etc.), cars and other motor vehicles, and capital goods is virtually non-existent in any Pacific Island country.

**Table 2:** Pacific ACP - countries - formal and informal employment

| | | | | | | formal sector employment (2004) | | | | | |
|---|---|---|---|---|---|---|---|---|---|---|---|
| 1 | 2 | 3 | 4 | 5 | 6 | 7 | 8 | 9 | 10 | 11 | 12 |
| Country | Sub-Region | Population | Population 0–14 years | Urban Population at last cencus | Urban Population | Working Age Population | Formal Sector Employment | % of 8 | **Fresh entrants of the workforce per year** | **Public Service** | **Informal Sector** |
| | | **(2013)** | **(% of total population 2013)** | **(%)** | **(Nos)** | **(15–64 years old)** | | | | **% of 8** | |
| Cook Islands | **Polynesia** | 15,200 | 28.5 | 72 | 10,916 | 7,276 | 5,900 | 81 | 341 | 22.0 | 1,376 |
| Niue | | 1,500 | 24.7 | 36 | 536 | n.a. | n.a. | n.a. | n.a. | n.a. | n.a. |
| Samoa | | 187,400 | 39.0 | 21 | 39,096 | 91,193 | 57,00 | 63 | 5,298 | 39.0 | 34,093 |
| Tonga | | 103,300 | 37.6 | 23 | 23,966 | 51,842 | 34,600 | 67 | 2,415 | 22.5 | 17,224 |
| Tuvalu | | 10,900 | 32.8 | 47 | 5,121 | n.a. | n.a. | n.a. | n.a. | 69.5 | n.a. |
| Fed. States of Micronesia | **Miconesia** | 103,000 | 33.8 | 22 | 22,688 | 61,786 | 15,137 | 24 | 3,008 | 52.5 | 46,649 |
| Kiribati | | 180,800 | 35.3 | 44 | 47,398 | 54,335 | 7,900 | 15 | 2,537 | 67.0 | 46,435 |
| Marshall Islands | | 54,200 | 40.2 | 65 | 35,314 | 29,634 | 10,300 | 35 | 1,604 | 48.0 | 19,334 |
| Nauru | | 10,500 | 38.9 | 100 | 10,500 | n.a. | n.a. | n.a. | n.a. | 47.0 | n.a. |
| Palau | | 17,800 | 19.9 | 77 | 13,569 | n.a. | n.a. | n.a. | n.a. | n.a. | n.a. |

| | | | | | | | | | | | |
|---|---|---|---|---|---|---|---|---|---|---|---|
| Fiji Islands | **Melanesia** | 859,200 | 28.2 | 51 | 435,973 | 487,450 | 120,000 | 25 | 18,600 | 38.0 | 367,450 |
| Papua New Guinea | | 7,398,500 | 39.4 | 13 | 961,805 | 3,320,217 | 187,320 | 6 | 159,618 | 40.0 | 3,132,897 |
| Solomon Islands | | 610,800 | 39.3 | 20 | 120,531 | 239,362 | 22,177 | 9 | 13,320 | 18.0 | 217,185 |
| Vanuatu | | 264,700 | 36.9 | 24 | 64,692 | 110,976 | 16,300 | 15 | 6,359 | 30.5 | 94,676 |
| Timor Leste | | 1,066,409 | 43.2 | 28 | 298,595 | 520,265 | 36874 | 7 | n.a. | 47.2 | 483,391 |

Sources: Population (Forum Countries) : SPC Pacific Islands Populations - Estimates the projections of demographic indicators of selected years, Sept. 2013, http://www.spc.int/sdd/ on Timor Leste: http://www.statistics.gov.tl/ Employment Data: World Bank ([2006]), p. 37, Table two. Six, Fresh entrants into the workforce pert year: PIFS 2006, Table four.

## Weber

Weber *Bandung: Journal of the Global South* 2015 **2**:6, doi: 10.1186/s40728-014-0009-z

In all Pacific Island countries semi-subsistence agriculture and other rural and urban informal sectors play a dominant role. Formal sector employment is lowest in Melanesian countries and Timor Leste where the informal sector consists of around 75 percent (Fiji) to 94 percent (Papua New Guinea) of the workforce. In Micronesia the informal sector is between 65 percent (Marshall Islands) and 85 percent (Kiribati) while the corresponding figures for Polynesia are between 19 percent (Cook Islands) and 37 percent (Samoa) (Table 2; Booth et al. [2006], World Bank [2006]).

Every year a number of young people wish to enter the workforce which is much bigger than jobs in the formal sector are available. Data gaps make it difficult to exactly quantify the number of missing formal employment. Alone in PNG year by year the annual net addition to the labour market is almost 160,000; in Fiji around 19,000 people join the labour market in a year (Table 2). For the entire Pacific Island region a very rough estimate is well above 200,000 – a lot if one considers that the formal economy in Pacific Island countries employs not much more than 500,000 (World Bank[2006]).

Many Pacific Island countries have started to scale down their public sectors. Chand ([2006]) highlights that the role of the public sector is more or less restricted to create an enabling environment for the expansion of private sector activities. In some Pacific Island countries tourism has become a major economic sector, but so far the potentials are not yet fully utilized. Direct employment in resorts, hotels, restaurants, and transportation as well as spill-over effects to other economic sectors is seen by governments as a very promising outlook for the future.

## Policies and Agreements that Affect Skilled Labour Mobility

Removing obstacles to the movement of people (as well as goods and services) is increasingly considered a way of contributing to economic development through meeting identified labour shortages, transferring skills, and increasing remittances (ADB [2008]).

Pacific Island countries greatly vary in their access to international migration opportunities for work purposes. Some have fairly open access to working opportunities abroad due to special bilateral agreements (e.g. between New Zealand and certain Polynesian states) or some states in the north-western Pacific which have Compact of Free Association (COFA) status with the USA (Keitner and Michael [2003]). People in other Pacific Island countries are more restricted in their by the immigration policies of destination countries (ACP-Observatory on Migration [2012]; see also Table 3).

**Table 3:** Migration data and schemes for Pacific Island countries and territories

| Country | Sub-Region | Population (2013) | Stock of Emigrants (in '000) | Stock of emigrants as % of population | Pacific Access Category (PAC) / Samoa Quota (SQ) (annual quota) | Unskilled Labour Schemes | Migration Opportunities |
|---|---|---|---|---|---|---|---|
| Cook Islands | **Polynesia** | 15,200 | 28.4* | 61.4%* | | | Full access to NZ |
| Niue | | 1,500 | 9.2* | 75.6%* | | | Full access to NZ |
| Samoa | | 187,400 | 120.4 | 67.3% | 1100 (SQ) | RSE, PSWPS2, SWP | |
| Tonga | | 103,300 | 47.4 | 45.4% | 250 | RSE, PSWPS1, SWP | |
| Tuvalu | | 10,900 | n.a. | n.a. | 75 | RSE, PSWPS2, SWP | Special deal with German merchant marine |
| Fed. States of Micronesia | **Miconesia** | 103,000 | 21.9 | 19.7% | | | Full access to USA (COFA) |
| Kiribati | | 180,800 | 6.4 | 6.5% | 75 | RSE, PSWPS1, SWP | Special deal with German/ Japanese merchant marine |
| Marshall Islands | | 54,200 | 10.5 | 16.6% | | | Full access to USA (COFA) |
| Nauru | | 10,500 | n.a. | n.a. | | PSWPS2, SWP | |
| Palau | | 17,800 | 8.0 | 38.8% | | | Full access to USA (COFA) |

| | | | | | | | |
|---|---|---|---|---|---|---|---|
| Fiji Islands | **Melanesia** | 859,200 | 182.2 | 21.3% | ** | RSE** | |
| Papua New Guinea | | 7,398,500 | 61.2 | 0.9% | | PSWPS1, SWP | |
| Solomon Islands | | 610,800 | 5.4 | 1.0% | | PSWPS2, SWP | |
| Vanuatu | | 264,700 | 3.9 | 1.6% | | RSE, PSWPS1, SWP | |
| Timor Leste | | 1,066,409 | n.a. | n.a. | | SWP | |
| Territory | | | | | | | |
| American Samoa | Poly | 66,692 | 45.7 | 66.7% | | | Full access to USA |
| French Polynesia | Poly | 271,831 | 4.2 | 1.5% | | | Full access to France |
| | | | | | | | (incl. French Pacific territories) |
| Tokelau | Poly | 1,162 | 2.9* | 65.3%* | | | Full access to NZ |
| Wallis & Futuma | Poly | 13,193 | n.a. | n.a. | | | (incl. French Pacific territories) |
| Guam | Micro | 192,090 | 48.7 | 27.1% | | | Full access to USA |
| Northern Mariana Islands | Micro | 63,517 | 8.9 | 10.1% | | | Full access to USA |

| | | | | | | | |
|---|---|---|---|---|---|---|---|
| New Caledonia | Mela | 252,331 | 2.9 | 0.9% | | | Full access to France |
| | | | | | | | |
| | | | | | | | |
| | | | | | | | |
| | | | | | | | |
| | | | | | | | (incl. French Pacific territories) |

Sources: for Population Data: SPC Population Data 2011 and Time Series from 1900; Migration rates and opportunity: Ware 200, information of emigrants: World Bank 2012 (except * Appleyard anf Stahl 1995); n.a. (not available) New Zealand government announced the "Recognized Seasonal Employer" (RSE) policy in October 2006 and it began in April 2007. Under RSE there are up to 8000 places available per year. Australia launched the "Pacific Islands Seasonal Workers Pilot Scheme" (PSWPS1) in Aug 2008, in September 2011 more countries were added (PSWPS2). The scheme ended in June 2012 and has been replaced by the Seasonal Worker Program (SWP), in which started on July 1, 2012 providing up to 12,000 places horticulture, accommodation providers, aquaculture venture and cotton and cane growers https://employment.gov.au/seasonal-worker-program .

** excluded for political reasons between 2007 and 2014.

Weber

Weber *Bandung: Journal of the Global South* 2015 **2**:6, doi:10.1186/ s40728-014-0009-z

When the WTO took up its work on January 1, 1995 a new era started for trade between Developed and Developing Countries. By signing the Marrakesh agreement of the Uruguay Round of GATT a General Agreement on Trade in Services (GATS) was established for the first time. The provision of GATS (Mode 4) allows temporary labour mobility for the purpose of service delivery.

Six of the Pacific Island countries (Fiji, PNG, Samoa, Solomon Islands, Tonga, and Vanuatu) are members of the World Trade Organization (WTO) and covered by its principles and agreements. The only Asian country amongst the Pacific ACP members, Timor Leste, hopes that becoming a member of ASEAN will help to boost economic progress and make labour movement to and from this young Asian country easier when skilled labour mobility within the ASEAN Economic Community is expected to become a reality by 2015 when the Asian Economic Community (AEC) is established (Ibrahim [2007]).

The WTO undertook much effort to regulate preferential trade agreements that before the establishment of the WTO had favoured some countries and disadvantaged others. Pacific Island countries enjoyed preferential trade with the European Union through the various Lomé Conventions (Grynberg [1998]). With the Cotonou Agreement of 2000 a phasing out of preferential trade started which was to be concluded through Economic Partnership Agreements (EPAs) between the EU and ACP member states at the latest by the end of 2007 (Rampa [2007]).

To become conform to WTO rules these EPAs need to cover *substantially all the trade* between the partners (Scollay and Grynberg [2005]). While trade under the Lomé Conventions the EU granted duty-free access of certain goods from ACP countries to the EU market the EPAs now require ACP countries to follow the principle of reciprocity in trade, which means that goods produced in the EU enjoy duty-free / duty-reduced access to the markets of ACP countries (Kimunguyi [2005]). Trade between the Least Developed Countries (LDCs) within the ACP and the EU can continue to be non-reciprocal under the Everything but Arms (EBA) initiative. For the Pacific Island countries this includes Kiribati, Samoa, Solomon Islands, Timor Leste, Tuvalu, and Vanuatu ([EU 2013]).

Negotiations are under way with the European Union for comprehensive Economic Partnership Agreement (EPA) with Pacific ACP countries (Kelsey [2007]). Fiji and Papua New Guinea have already interim EPAs which were signed in 2009 and ratified in early 2011 (EU [2014]). Overall the progress in achieving a comprehensive EPA of all Pacific ACP countries[c] has been slow and burdensome. In December 2013 Fiji withdrew from the Pacific ACP trade talks organised by the Forum Secretariat maintaining the need to hold a regional meeting first without the EU's presence or pressure from the Forum Secretariat.

[c]Timor Leste is not included in the Economic Partnership Agreement, but it has an observer status and most likely would be able to join once the agreement is ready to be signed.

Developing countries also have preferential market access under various Generalized System of Preferences (GSP) (e.g. the EU, USA, Australia, and New Zealand have such systems) that allow preferential market access, e.g. reduced tariff rates for goods from developing countries (Herz and Wagner [2011]). The EU has been revising its GSP, excluding a number of countries that earlier had been covered (e.g. Fiji and PNG, which now fall under Economic Partnership Agreement Market Access Regulations, [EU 2013]). From the Pacific Island countries Cook Islands, Marshall Islands, Federated States of Micronesia, Nauru, Niue, and Tonga have market access to the EU through the GSP ([EU

2013]). The EU also has introduced a GSP + facility, which provides additional preferences to countries which ratify and implement 27 international conventions on core human and labour rights, the environment and good governance (Verma and Elman [2007], Switzer [2008]). None of the 15 Pacific ACP countries at the moment qualifies for GSP + preferential market access to the EU ([EU 2013a]).

Pacific Island countries have negotiated (and are right now negotiating) a variety bilateral, regional and international trade agreements. The oldest of these agreements is the South Pacific Regional Trade and Economic Cooperation Agreement (SPARTECA), which started in 1981 and granted Forum Island countries preferential (non-reciprocal) access to Australia and New Zealand for a specified selection of goods originating from FICs (Gold and Thakur [1982], Rolfe [2007]). The Pacific Island Countries Trade Agreement (PICTA) between Forum Island countries (except Australia and New Zealand) entered into force in 2003. In 2012 a Trade in Services agreement was added to PICTA (PICTA TIS), which also has the provision of a Temporary Movement of Natural Persons (TMNP) protocol (Mahadevan and Asafu-Adjaye, [2013]). The TMNP protocol however is not yet open to signature.

Another regional trade agreement is the Melanesian Spearhead Group (MSG) Trade Agreement between Papua New Guinea (PNG), Solomon Islands, Vanuatu and Fiji), which started in 1993. In 2012 the MSG also introduced a Skills Movement Scheme (SMS) (MSG [2012]).

At international levels agreements include the Pacific Agreement on Closer Economic Relations (PACER), which includes the Pacific Island countries, Australia and New Zealand (Gathii [2011], Kelsey [2006]). PACER actually was not designed as a Free Trade Agreement, but as a safety valve for Australia and New Zealand for the case that any Forum Island Country (FIC) commences negotiations on Free Trade Agreements with other developed countries (e.g. with the EU). Such negotiations then was to trigger the need for consultations with Australia and New Zealand and negotiations for a Free Trade Agreement between those FICs who commenced negotiations for free trade agreements with other developed countries and Australia and New Zealand (Gathii [2011], Gupta [2008]). Besides this PACER stipulated that negotiations between the FICs and Australia and New Zealand for a Free Trade Agreement needed to start not more than eight years after PICTA came into force (Kelsey [2004]). Negotiations for PACER-Plus, the proposed free trade agreement between the Forum Island Countries and Australia and New Zealand, were launched at the Pacific Islands Forum Leaders meeting in August 2009 (PIFS [2009]).

In all the agreements that are currently under negotiation Pacific Island countries are eager to include regulations about labour mobility (GATS Mode 4 and beyond). Under PACER-Plus they hope that unilaterally established (un-skilled) temporary labour mobility schemes by Australia and New Zealand can be made more sustainable and binding, if PACER-Plus includes aspects of temporary labour mobility. Chances seem to be slim that this can be achieved. There is little indication that new and binding access to Australian and New Zealand labour markets will result from the talks – a prospect that would leave many island nations without a key incentive to remain at the negotiating table (Morgan [2012]; PIFS [2009]).

In the 2010 and 2011 annual reports of the Office of the Chief Trade Adviser (OCTA) Regional Labour Mobility (beyond Mode 4) is listed as a priority agreed by Forum Trade Ministers in October 2009. On its website OCTA stresses that for many FICs, regional labour mobility (beyond Mode 4) is the key, if not the most important, issue in PACER Plus negotiations. Many Pacific Island Countries think that PACER Plus will not lead to mutually beneficial and development enhancing economic integration unless it results in greater integration of labour markets in the region. Australia has recognised the importance of labour mobility, but maintains it is a complex area of negotiations (http://www.octapic.org/pacer-plus-nearing-conclusion-says-chief-trade-adviser/ ). Pacific leaders also hope that PACER-Plus increases labour mobility between Island countries (Fiji Times, 17 March 2013).

Pacific ACP countries made a similar experience in the negotiations for EPAs (or a comprehensive EPA between the EU and Pacific ACP countries). The PACPS are targeting an annual quota of 10,000 semi-skilled and intermediate trainees (IT) from the PACPS to work in EU member countries under the EPA (PANG [2007]). The EU is very hesitant to include the mobility of temporary labour in the EPA negotiations (Manservisi and Falkenberg [2006]; Rampa [2007]).

## Analysis of Skilled and Unskilled Labour Mobility from Pacific Island Countries

Faist and Fauser ([2011]) highlight important changes in migration paradigms: they notice the desirability of temporary labour migration based on the expectation that temporary migrants will constitute no loss in human capital and furthermore transmit a higher percentage of their income than permanent immigrants. Such temporary labour migration can also help to spread the benefits of migration widely and create familiarities with destination countries. Temporary labour migration can help Pacific Islanders acquire skills, knowledge and insights that can become crucial when they later have to leave their home countries for good.

Stahl and Appleyard ([2007]) hint at the development aspects of programmes by New Zealand and Australia to grant permanent access to people from Pacific Island countries for work purposes. New Zealand has granted citizenship to the people of the Cook Islands, Niue and Tokelau. The country also has a quota for Samoan citizen under which 1,100 Samoan nationals annually can become permanent residents. This Samoan quota exists since 1970. The Pacific Access Category (PAC) of 2002 allows 250 persons from Fiji[d], 250 from Tonga, 75 from Kiribati and 75 from Tuvalu to become permanent citizens. Under the SQ and PAC people from Pacific Island countries, who would not qualify under the general programs for skilled immigrants as their skills levels are often too low can enter New Zealand (see also Table 3).

[d]Fiji joined the scheme in 2003, and after the coup of 2006, Fiji's participation in the scheme discontinued.

Australia and New Zealand have also bilateral agreements with a number of Pacific Island countries in the area of seasonal migration of unskilled labour, such as the Recognised Seasonal Employer (RSE) policy New Zealand introduced in 2007 and a similar scheme Australia launched a year later. Such schemes had existed in the late 1960s for Fijian, Tongan and Samoans to work mainly in agriculture and forestry until the first coup in Fiji (1987) brought the scheme to an end (Lee [2009]:10). Under the RSE policy, at present, up to 8,000 persons from (in order of absolute importance) Vanuatu, Tonga, Samoa, the Solomon Islands, Tuvalu and Kiribati can come to New Zealand for seasonal agricultural work. The Australian scheme allows people from Papua New Guinea, Vanuatu, Kiribati, Tonga, Samoa, the Solomon Islands, Tuvalu, and (since 2012) Timor Leste to enter Australia for seasonal unskilled employment mainly in the agricultural sector. Other economic sectors showed interest in employing temporary workers from the Pacific such as the New Zealand dairy and meat processing industries.

In the extension of the Australian scheme in 2012 a three year trial for labour recruitment to the tourism industry has been announced. Cotton and sugar cane growers as well as fisheries have become part of this trial which provides 1,500 jobs. Gibson and McKenzie ([2011]) suggest that the New Zealand RSE can make a meaningful development contribution to Pacific Island nations. Walmsley et al. ([2009]) argue that Australia and New Zealand would gain considerably from increasing quotas, particularly on unskilled labour and also the Pacific Island economies gained a lot from sending unskilled labour to Australia and New Zealand.

Initiatives taken up by metropolitan countries or by Pacific Island countries on labour mobility are highly controversial. Some see in them a return to the nineteenth century when colonial labour trade was (mis) used to develop the Australian economy. Others see them as a way to bring development to Pacific Island countries. Opening Pacific Islands' economies even more to imports from Australia and other countries to the disadvantage of their own, rather fragile and vulnerable economies might be easier to bear when labour mobility becomes a binding part of trade agreements.

However, it is very unlikely that Australia or New Zealand will give away the right to regulate labour movements considering benefits and potential threats to their own economies.

Larger Pacific Island countries, in particular Fiji and Papua New Guinea, train big numbers of professionals and graduates year by year. Many leave, making their skills available to others. The skills movement scheme of the MSG tries to balance labour supply and demand within the four economies of the MSG group. However, such schemes can only be successful if a bigger number of skilled labourers that are willing and able to migrate chose to go to a Pacific Island country rather than migrate to a metropolitan country. To make this happen requires incentives—incentives in terms of attractive wages, working conditions and living qualities. It also requires the establishment of a

Pacific identity, where it becomes a value in itself to live and work in a country that culturally is similar to where migrants come from.

Such considerations also become relevant when places for people need to be found that leave their homes because of climate change. It is now the time to equip people with skills that enable them to positively contribute to the societies that are supposed to receive them. It is equally important that there is a Pacific Island identity that sees it natural to help out and welcome those who need support, who need new homes.

## Environmentally-Induced Migration in the Pacific Islands

Pacific Islands have figured prominently in discourses about environmentally induced migration. Sea-level rise, the increase in intensity of natural hazards, the destruction of coral reefs and many other aspects of climate change lead to changes in the environment. At the same time they are social events: they create spaces of consideration, of discourses, of fear and assumption about an unsure and an unknown future. This is, in particular, true for the fate of those who live in places that might be submerged sometime in the future.

It is obvious that changes in environmental quality can make people leave their homes. Still the terminology "environmental refugees" is rejected by many. Legal frameworks about refugees don't accept environmental reasons when granting refugee status.

There is little doubt that Pacific Island countries are much exposed to the impacts of climate change. There is also little doubt that many Pacific Island countries have been a source region for big numbers of emigrants. The question is, if there are indications that climate change and migration in the Pacific have been related in the past and if for the future one can expect big numbers of people from Pacific Island countries who are driven away from their homes.

The search for well-being and security is the major reason why people migrate (Gasper [2011]). Hau'ofa ([1993]) described the pre-colonial Pacific as a 'sea of islands' within which people moved freely and frequently, created social networks, traded and exchanged goods, and at times engaged in conflict and attempted to exert dominance over one another (cited in Lee [2009]: 1). Mobility within the Pacific Islands region created pluralistic, multicultural, transnational societies that benefited much from the mixing of people.

Small Island States are amongst those countries that are most exposed to various impacts of climate change. They "have characteristics which make them especially vulnerable to the effects of climate change, sea-level rise and extreme events" (IPCC, [2007]: 15; IPCC [2012]: 184). Their high vulnerability is intensified by low adaptation capacities and capabilities.

The main climate change challenges in the Pacific Islands are flooding, tropical storms (cyclones and hurricanes) and drought. Sea-level rise as a slow inset, low intensity hazard can make whole atoll island states disappear. Such hazards and risks, environmental, economic and social, and climate change and its impacts are major threats to development achievements (Asia-Pacific Disaster Report [2010]). Within a few decades these island states could become uninhabitable and people have to move.

Campbell et al. ([2010]) have identified 86 cases of relocation in Pacific Islands between 1920 and 2000. 50 of them are related to environmental conditions. In the very near future resettlement most likely will increase tremendously. In February 2014 Fiji's President affirmed to the people of Kiribati "that you will not be refugees [...]. You will be able to migrate with dignity [...]. If the sea level continues to rise because the international community won't tackle global warming, some or all people of Kiribati may have to come and live in Fiji" (President Nailatikau quoted in Fiji Sun, Feb. 14, 2014).

End of May 2014 the Government of Kiribati purchased 5,460 acres of land on Vanua Levu, Fiji's second biggest island for AUS$9.3 million (Fiji Broadcast Corporation, May 29th, 2014)[e]. Although President Tong stressed that the land purchased is to strengthen Kiribati's food security, and not for relocation of I-Kiribati people, there are many who suggest that this was only done to contain the political impact of such move back home in Kiribati.

[e]5,460 acres (22.1 $km^2$) is considerably bigger than South Tarawa (15.76 $km^2$) where almost half of Kiribati's population is living.

In Fiji a big number of relocation activities have already started. In January 2014 Fiji's Prime Minister announced that 676 communities have been identified that are threatened by sea-level rise generated by climate change; 42 of them shall be relocated within the next 10 years (Nailatikau[2014]).

## Of Good and Bad Migrants

People living in the Pacific Islands are not isolated from events elsewhere. Through media and their own experience they learn that there are different kinds of migrants, and that in not all of the cases migration brings the outcomes people hope for. There always have been asymmetries when it came to various forms of mobility.

Labour immigration is vital to many countries. Countries like Australia, New Zealand, Canada and the USA are built on migration, on *'good immigrants'*. Refugees however are often perceived as *'bad migrants"*, as people who enter countries illegally, misuse social security and welfare provisions of these countries and threaten the well-being and even physical security of the law-abiding citizens of these countries. Often the public does not realise (or admit) that these people face prosecution, are displaced and depend on support from outside. It seems that the legal protection of refugees adds rigour to the attempt of countries to keep their borders tight and to prevent that such people even reach the territory of countries to seek the status of refugee. If they do so then there usually is a long process waiting for asylum seeker, which often deprive them of basic rights.

There is little reason to assume that this will be different when climate-change refugees from Pacific Islands are knocking on the doors of their powerful neighbours.

In the Pacific in particular Australia has become an example of relentless refugee policies which many see as a violation of the relevant conventions for refugees and human rights. People living in countries that soon might disappear from the world map are frightened that then they will be exposed to inhumane treatment like today asylum seeker in Australian detention centres in Papua New Guinea and Nauru: locked away for indefinite time, with few rights and even less opportunities to decide over their lives. Keeping this in mind it looks sarcastic that life-styles like those in Australia seem to be the most important cause for climate change, for the misery many Pacific Islanders will suffer in a few decades from now despite having contributed nothing to it.

Knowing that in a few decades they might have to look for places where they can live Pacific Islanders are not eager to become environmental refugees, even if this would become legally possible. Leaders of Pacific Island countries emphasise again and again that they do not want their people to be seen as potential environmental refugees. Kiribati President Anote Tong, for example, does not leave any doubt that he is aware that a time might come when the 32 atolls of his country are no more unsuitable to provide a home to the people of Kiribati. President Tong has started to find alternatives for more than 100,000 people, but not as environmental refugees, but to allow them to *'migrate in dignity'*.

Considering the hostile attitude towards refugees it seems prudent to work towards solutions within Pacific Islands that enhances South-South cooperation and solidarity and resists to new dependencies towards former colonial masters. Metropolitan powers however have a role to play: they should meet their obligations in helping Pacific Island countries to generate capacities so that people can contribute to the societies of the countries receiving them.

# CONCLUSIONS

Pacific Islanders have been mobile since people started settling in this part of the world. They always contributed to the betterment of societies they received tham. The near future will show, if a new stage of migration becomes dominant: the migration of people who have to leave their homes because of climate change. If this becomes reality then much has to be learnt from historical examples of migration, in particular those, where asymmetries of power resulted in dependencies and exploitation as a result of migration. To make people climate change refugees will compromise their right to live in dignity. All over the world migrants who can contribute to the societies of destination countries are less endangered to become socially isolated and not welcome.

The challenge is best taken up amongst the Pacific Island countries. Solidarity based on similar histories and cultural traits should make this easier to happen. Many Pacific Island countries already have meaningful populations of other Pacific Island countries. Challenges arising from climate and environmental change have the potential to strengthen Pacific Island identity and can show that the people in the Pacific Islands are able to master these challenges. Still financial support from metropolitan countries will be required to strengthen capacities in the Pacific Islands. Such support should not create new dependencies and increase existing ones, but should be based on the responsibilities the countries at the Pacific Rim have as the contributed to the challenge of climate change through their degree of industrialization and unsustainable lifestyles as major sources of emissions of greenhouse gases.

Today we do not even know the number of (additional) migrants because of climate change. In the face of slow environmental change, those who are able to move—those with money, social networks and alternative livelihoods—may migrate independently. The vulnerable poor, those with no capacity to move when environments deteriorate, the young and elderly may be left behind or forced to resettle later (Warner and Laczko [2008]: 60). These then will become climate-change refugees: with few capacities and skills, little wanted by countries that should receive them, dependent on support and benevolence, and endangered to become a pawn in political in-fighting.

There is even the possibility of people being left behind in adverse environments. Such fear is not baseless: the majority of people in the so-called developed world bother little when people (already now) are living under extremely adverse environmental and/or social conditions. Evacuating (Kelman[2008]) sinking islands in the Pacific most likely will not happen. It is more likely that the last plane and ship leaves Kiribati (or Tuvalu) and people are still there, not forgotten, but never intended to be brought to safety, neither socially nor environmentally.

# REFERENCES

1. 2012. Assessment of South-South Labour Mobility in the Pacific region, Terms of Reference for the call for tenders for undertaking research commissioned by the ACP Observatory on Migration. ACP Observatory on Migration, Brussels.
2. 2008. Skilling the Pacific. Technical and Vocational Education and Training in the Pacific. Manila, Asian Development Bank.

3. 2010. Protecting Development Gains, Reducing Disaster Vulnerability and Building Resilience in Asia and the Pacific. ESCAP / UNISDR, Bangkok, Thailand.
4. Booth, Heather, Guangyu Zhang, Maheshra Rao, Fakavae Taomia, and Ron Duncan. 2006. Population pressures in Papua New Guinea, the Pacific Island Economies, and Timor Leste. Working Papers in Demography No. 102, Research School of Social Sciences, Canberra: The Australian National University.
5. Campbell, JR. 2014. Climate-Change Migration in the Pacific. The Contemporary Pacific 26 (1): 1–28.
6. Campbell, JR, and R Bedford. 2014. Migration and Climate Change in Oceania. In People on the Move in a Changing Climate. The Regional Impact of Environmental Change on Migration, ed. Piguet E, Laczko F, Springer, Dordrecht Heidelberg New York London.
7. Campbell, JR, M Goldsmith, and K Koshy. 2010. Community Relocation as an option for adaptation to the effects of climate change and climate variability in Pacific Island Countries (PICs). Final report for APN project 2005-14-NSY-Campbell. University of Waikato, Hamilton.
8. Chand, S. 2006. Working Paper 06–08. The Australian National University, Crawford School of Economics and Government, Canberra.
9. Duncan, R, and C Voigt-Graf. 2008. Labour market scenarios for the Asian Decent Work Decade in the Pacific Island countries. International Labour Organisation Asia-Pacific Working Paper Series. ILO Regional Office for Asia and the Pacific, Manila.
10. EU 2014. Overview of EPA negotiations. Updated 7 February 2014. Available from http://trade.ec.europa.eu/doclib/docs/2009/september/tradoc_144912.pdf.
11. EU 2013. Revised EU trade scheme to help developing countries applies on 1 January 2014. Press Release, December 19. Available from http://europa.eu/rapid/press-release_MEMO-13-1187_en.htm?locale=FR.
12. EU 2013a. 10 countries to benefit from EU preferential trade scheme GSP+as of 1 January 2014. Press Release, December 30. Available from http://trade.ec.europa.eu/doclib/press/index.cfm?id=1006

13. Faist, T, and M Fauser. 2011. The migration-development nexus: towards a transnational perspective. In The migration-development nexus: a transnational perspective, ed. Thomas F, Margit F, Peter K, Palgrave Macmillan, Basingstroke.
14. Firth, S. 1978. German Labour Policy in Nauru and Angaur, 1906–1914. The Journal of Pacific History 13 (1): 36–52.
15. Flanagan, Tracey, Meredith Wilkie, and Susanna Iuliano. 2003. Australian South Sea Islanders A century of race discrimination under Australian law. The role of South Sea Islanders in Australia's economic development Available from https://www.humanrights.gov.au/erace-archives-australian-south-sea-islanders
16. Gasper, Des. 2011. International migration, well-being and transnational ethics. In Transnational migration and human security, edited by Truong, T.-D. and Gasper, D. Heidelberg, Dordrecht, London, New York: Springer.
17. Gathii, JT. 2011. The neoliberal turn in regional trade agreements. Washington Law Review 86 (3): 421–474.
18. Gibson, J, and D McKenzie. 2011. Australia's Pacific Seasonal Worker Pilot Scheme (PSWPS): development impacts in the first two years. Department of Economics, Working Paper in Economics, 09/11. University of Waikato, Hamilton.
19. Gold, H, and R Thakur. 1982. New Zealand and Australia: Free Trade Agreement Mark II. The World Today 38 (10): 402–410.
20. Goss, J, and B Lindquist. 2000. Placing Movers: An Overview of the Asian-Pacific Migration System. The Contemporary Pacific 12 (2): 385–414.
21. Grynberg, R. 1998. The WTO incompatibility of the Lomé trade provisions. Asia Pacific School of Economics and Management, Working Papers 98/3. The Australian National University, Canberra.
22. Gupta, S. 2008. Changing Faces of International Trade: Multilateralism to Regionalism. Journal of International Commercial Law and Technology 3 (4): 260–273.
23. Hau'ofa, E. 1993. A New Oceania: Rediscovery of our Sea of Islands. University of the South Pacific, Suva, Fiji.

24. Herz, B, and M Wagner. 2011. The Dark Side of the Generalized System of Preferences. Review of International Economics 19 (4): 763–775.
25. Hills, HL. 1984. Compact of Free Association for Micronesia: Constitutional and International Law Issues. The International Lawyer 18: 583–608.
26. Ibrahim, Fatimah Wati. 2007. Assessing regional integration: The progress of ASEAN regional grouping. In: International Economic Conference on Trade and Industry (IECTI) 2007, 3–5 December 2007, Bayview Hotel Georgetown, Penang. Available from http://repo.uum.edu.my/2412/
27. 2007. Climate Change 2007: Impacts, Adaptation and Vulnerability. Contribution of Working Group II to the Fourth Assessment Report of the Intergovernmental Panel on Climate Change Paul P, Osvaldo C, Jean P, Linden P, Clair H Cambridge University Press, Cambridge, UK.
28. Intergovernmental Panel on Climate Change (IPCC) 2012. Managing the Risks of Extreme Events and Disasters to Advance Climate Change Adaptation. A Special Report of Working Groups I and II of the Intergovernmental Panel on Climate Change [Field, C.B., V. Barros, T.F. Stocker, D. Qin, D.J. Dokken, K.L. Ebi, M.D. Mastrandrea, K.J. Mach, G.-K. Plattner, S.K. Allen, M. Tignor, and P.M. Midgley (eds.)]. Cambridge University Press, Cambridge, UK, and New York, NY, USA, 582 pp.
29. Jupp, J. 2007. From White Australia to Woomera. The story of Australian immigration. Cambridge University Press, Port Melbourne.
30. Keitner, CI, and W Michael Reisman. 2003. Free Association: The United States Experience. Texas International Law Journal 1 (39): 1–63.
31. Kelman, I. 2008. Island evacuation. Forced Migration Review, Issue 31: 20–21.
32. Kelsey, J. 2004. World trade and small nations in the South Pacific region. Kansas Journal of Law and Public Policy 14: 247–306.
33. Kelsey, J. 2006. Free trade agreements: boon or bane?: through the lens of PACER. Victoria University of Wellington Law Review 37: 391–419.

34. Kelsey, J. 2007. Going Nowhere in a Hurry? The Pacific's EPA Negotiations with the European Union. Victoria University of Wellington Law Review 38: 81–103.

35. Kimunguyi, Patrick. 2005. The European Union and Developing Countries the Challenges of Trade Liberalisation in the Contonou Process, Paper of Contemporary Europe Research Centre, University of Melbourne. Available from http://www.eusa-japan.org/download/eusa_ap/paper_PatrikKimunguyi.pdf.

36. Lee, H. 2009. Pacific Migration and Transnationalism: Historical Perspectives. In Migration and Transnationalism. Pacific Perspectives, ed. Helen L, Steve Tupai F, ANU Press, Canberra.

37. Mahadevan, R, and J Asafu-Adjaye. 2013. Unilateral Liberalisation or Trade Agreements: Which Way Forward for the Pacific? The World Economy 36 (10): 1355–1372.

38. Manservisi, Stefano, and Karl Falkenberg. 2006. Leaked letter from the EC's Falkenberg and Manservisi to Tavola on Pacific EPA Available from http://www.bilaterals.org/IMG/pdf/EC_Falkenberg_Manservisi_to_Tavola_Oct-06.pdf.

39. Matisoo-Smith, E, and JH Robins. 2004. Origins and dispersals of Pacific peoples: Evidence from mtDNA phylogenies of the Pacific rat. Proceedings of the National Academy of Science USA 101 (24): 9167–9172.

40. 1990. Labour in the South Pacific. James Cook University of Northern Queensland, Townsville, Qld.

41. Morgan, Wesley. 2012. Labour mobility through PACER-Plus? An unlikely prospect. Devpolicy Blog from the Development Policy Centre. Available from http://devpolicy.org/labour-mobility-through-pacer-plus-an-unlikely-prospect20120402/.

42. MSG (Melanesian Spearhead Group) 2012. MSG Skills Movement Scheme. Available from http://www.msgsec.info/index.php/msg-skills-movement-scheme

43. Munro, D. 1990. Origins of Labourers in the South Pacific: commentary and statistics. In Labour in the South Pacific, edited by Clive Moore, Jacqueline Leckie, and Doug Munro. James Cook University of Northern Queensland, Townsville, Qld.

44. Nailatikau, Ratu E. (2014). His Excellency Ratu Epeli Nailatikau- Address at a Function on His State Visit to Nauru.

Available from http://updates7131.rssing.com/browser.php?indx=5440860&item=61.

45. 2007. Social Impact Assessment of the Economic Partnership Agreement (EPA) being negotiated between the European Community and the Pacific ACP states'. Report to the Pacific Islands Forum Secretariat. Suva, PANG.
46. PIFS (Pacific Islands Forum Secretariat) 2009. Fourteenth Pacific Islands Forum, Cairns, Australia, 5 – 6 August 2009, Forum Communiqué. Available from http://www.forumsec.org/resources/uploads/attachments/documents/2009%20Forum%20Communique,%20Cairns,%20Australia%205-6%20Aug.pdf.
47. Rampa, Francesco. 2007. Implementation of Article 37(4) of the Cotonou Agreement. Provision of technical support to assist the Pacific ACP Region in the Review of EPA Negotiations. Draft Interim Report submitted to the Pacific Islands Forum Secretariat. Available from http://62.58.77.233/library/files/Rampa_EN_0107_ECDPM_Implementation-of-article-37-4-of-the-Cotonou-agreement.pdf.
48. Ridgell, R. 2006. Honolulu. Bess Press, Hawai'i.
49. Rolfe, J. 2007. Many Small States, two regions, different constructions. Social and Economic Studies 56 (1/2): 96–134.
50. Scollay, R, and R Roman Grynberg. 2005. Substantially All Trade: Which Definitions are fulfilled in Practice? An Empirical Investigation. Commonwealth Secretariat, London.
51. Shineberg, D. 1967. They Came for Sandalwood: A Study of the Sandalwood Trade in the South-West Pacific, 1830–1865. Melbourne University Press, Melbourne.
52. Stahl, Charles W., and Reginald Appleyard. 2007. Migration and development in the Pacific Islands: lessons from the New Zealand Experience. Canberra: Australian Agency for International Development (AusAID).
53. Switzer, S. 2008. Environmental Protection and the Generalized System of Preferences: A Legal and Appropriate Linkage? The International and Comparative Law Quarterly 57 (1): 113–147.
54. Verma, A, and G Elman. 2007. Labour Standards for a Fair Globalization for Workers of the World. The Good Society 16 (2): 57–64.

55. Walmsley, Terrie, Syud Amer Ahmed, and Christopher Parsons. 2009. The impact of liberalizing labour mobility in the Pacific Region. GTAP Working Papers No. 31. West Lafayette: Centre for Global Trade Analysis, Purdue University.
56. Ward, RG. 1972. The Pacific Beche-de-mer Trade with Special Reference to Fiji. In Man in the Pacific: Essays on Geographical Change in the Pacific Islands, ed. Ward RG, Clarendon, Oxford.
57. Ward, RG. 1989. Earth's empty quarter? The Pacific Islands in a Pacific Century. The Geographical Journal 155 (2): 235–246.
58. Warner, K, and F Laczko. 2008. A global research agenda. Climate change and displacement Forced Migration Review 31: 59–60.
59. 2006. At Home and Away – Expanding job opportunities for Pacific Islanders through labour mobility. The World Bank, Washington, D.C..
60. 2011. Migration and remittances factbook 2011. World Bank, Washington, DC.

Chapter 4

# Application of Active EM-Calcium in Green Agricultural Production – Case Study in Tomato and Flue-cured Tobacco Production

Xiaohou Shao[1,2], Tingting Chang1[,3], Maomao Hou[1,3], Yalu Shao[3], and Jingnan Chen[3]

[1]Key Laboratory of Efficient Irrigation-Drainage and Agricultural Soil-Water Environment in Southern China of Ministry of Education, Hohai University, Nanjing, PR China

[2]Nanjing Ning-ya Environmental Science and Technology Limited Company, Nanjing, PR China

[3]College of Water Conservancy and Hydropower Engineering, Hohai University, Nanjing, PR China

## INTRODUCTION

EM (Effective Microorganisms) technology has long been widely applied to agricultural production in China. EM contains more than 80 kinds of beneficial microorganisms, including yeast, lactic acid bacteria, actinomycetes and photosynthetic bacteria (Shao et al., 2013; Zhou et al., 2008). They can quickly decompose organic matter, metabolize antioxidant substances and inhibit the proliferation of harmful microorganisms (Higa, 1997). EM was supposed to have positive effects on reforming soil nematode community structure (Hu and Qi, 2013a), increasing crop yield and improving soil properties (Daly and Stewart, 1999; Khaliq et al., 2006).

High calcium content in plants has been associated with increased resistance to diseases (Berry et al., 1988), not merely the well-known tomato blossom-end rot (BER) (Sonneveld and Voogt, 2009), calcium nutrition significantly affects the resistance of tomato seedlings to the bacterial wilt caused by Ralstonia solanacearum (Yamazaki, 1995), other diseases were also reported to be suppressed by high calcium concentration in host plants (Almeida et al., 2009; Berry et al., 1988;Chiasson et al., 2005). BER is a major physiological disorder in tomato that creates up to 50% losses (Taylor, 2004), besides the inducing factors of high temperature, high salinity (Adams, 1993), BER has long been considered to be a Ca-deficiency disorder, and its incidence increases in cultivation at low Ca concentrations (Raleigh, 1944), however, most tomatoes contain little calcium ($Ca^{2+}$), uptake and translocation of cationic nutrients including $Ca^{2+}$in plants plays an essential role in the physiological processes (Chung et al., 2010). According to early studies, the exogenous calcium supplied through tomato roots (Hall, 1977; Sachan and Sharma, 1981) or through leaves (Eraslan et al., 2007; Freitas et al., 2012) by the spraying methods can both increase the Ca uptake of tomato plants and enhance the immunity to diseases.

Flue-cured tobaccos are important industrial crop significant to the national economy for China. In southwest China, most of the rain falls during the early and middle growth stages of flue-cured tobacco, and periodic drought often happens at the later growth period of flue-cured tobacco, this will not enable the upper leaves maturing normally, resulting in a poor availability of upper leaves, so it is important to discover technologies for the drought resisting and the tobacco quality and yield improvement (Hou et al., 2012; Hou et al, 2013). Water-retaining agent has been applied in many areas (Kumar and Dey, 2011; Truax and Gagnon, 1993). However, in studies of flue-cured tobacco cultivation, application of the water-retaining agent is still in a research vacant. At present, many studies have reported the application of EM in agricultural production (Hu and Qi, 2013a, b; Javaid, 2010; Khaliq et al., 2006). EM can quickly decompose organic matter, metabolize antioxidant substances and inhibit the proliferation of harmful microorganisms (Daly and Stewart, 1999; Daming, 1999; Heo et al., 2008).

In conclusion, many studies (Heo et al., 2008; Hu and Qi, 2013b; Khaliq et al., 2006) have been undertaken on the application of EM as base fertilizer or as a component of base fertilizer, however, there has been little published concerning using EM as foliage fertilizer, especially making EM into calcium nutrient solution. In recent years, our team invents a new product based on EM technology and names it as Active EM-Calcium. Active EM-Calcium is prepared by special fermentation process with lime and EM, which contains chelating calcium (Ca) and microorganisms in the solution. In order to verify the effect of Active EM-Calcium on tomato and flue-cured tobacco production, two researches were carried out from 2012 to 2013.

# EFFECTS OF EM-CALCIUM ON PRODUCTION OF GREEN HOUSE TOMATO

## Materials and Methods

### *Test Site*

The greenhouse experiments were carried out in plastic greenhouse during 2012 at the Vegetables and Flowers Institute of Nanjing (SW of China, lat. 31°43′ N, long. 118°46› E) in a well stirred heavy lay loam. The local climate is subtropical monsoon, with average annual rainfall of 1106.5mm, temperature of 15.7°C and humidity of 81%. Affected by the No. 1211 (international numbering) severe tropical storm "HAIKUI", temperature of experimental field in early August saw a drop of 10°C-14°C, mixed with heavy winds and rains, but due to the short continuance, greenhouse crops were little influenced by the storm. Average temperature during the experiment period was 21.9°C (May), 25.7°C (June), 29.8°C (July), 27.9°C (August), 22.3°C (September) respectively, which was 1.0°C, 0.9°C, 1.5°C, 0.6°C,-1.0°C higher than the same period in recent years. Peak temperature was 37.9°C recording on July 29$^{th}$. The experimental soil properties were as follows: organic matter 14.34 g $kg^{-1}$, available nitrogen 104.17 ppm, available phosphorus 26.48 ppm, available potassium 184.70 ppm.

## Experimental Design

Lab experiment: 28 g lime was accurately weighed in a 100 ml volumetric flask, then added DI water to dissolve and kept constant volume to 1000 ml, and the 2% $Ca^{2+}$suspension was available by shaking the mixture well. Then the prepared calcium suspension was mixed with different volume EM, molasses (EM and the molasses were supported by EMRO Limited Company, Nanjing Branch) and DI water to make up 5 treatments with three replications. The ingredients of treatment T1-T5 were displayed as Table 1. These sealed EM-Calcium mixtures would be fermented for 3-6 days in a orbital shaker with constant temperature, they were taken out and setted aside when the precipitates were basically dissolved. The fermented treatment with highest calcium solubility was diluted to keep the active EM-calcium solution with 2.0‰ $Ca^{2+}$concentration.

**Table 1:** The components of EM-calcium solution with different contents

| Treatment | **2% $Ca^{2+}$ suspension (ml)** | **EM (ml)** | **Molasses (ml)** | **DI water (ml)** | **Theoretical $Ca^{2+}$concentration** |
|---|---|---|---|---|---|
| T1 | 45 | 25 | 25 | 205 | 3.0‰ |
| T2 | 60 | 30 | 30 | 180 | 4.0‰ |
| T3 | 75 | 40 | 40 | 145 | 5.0‰ |
| T4 | 90 | 45 | 45 | 120 | 6.0‰ |
| T5 | 105 | 50 | 50 | 95 | 7.0‰ |

Field experiment: the six week old tomato seedlings (*"21$^{st}$ Century Crown"*) were transplanted to the experiment fields on June 7$^{th}$, conventional field management were carried out fairly among the treatments, no additional light, heat, or $CO_2$ were provided. The experimental field was ploughed several times and fertilized with 700kg hm$^{-2}$ compound fertilizer (N: $P_2O_5$: $K_2O$=1:2:2) in May. Irrigation systems for tomato were accorded with the local farming practices. At 8:30 every morning, micro sprayers were adopted to spray the active EM-calcium solutions 2 ml once in four days on different tomato organs. The treatments were set based on the sprayed organs, including spraying root, spraying flower, spraying leaves near the newborn fruits, spraying one week old fruit, spraying three week old fruit, and a control, hereafter referred as SR, SF, SL, SO, ST and CK respectively. A plastic film was used to keep apart the other parts when spraying one tomato organ.

## Measurements

When tomato was ripened, two fruits were harvested from one plant, and about 10 g tomato flesh per fruit was taken along the longitudinal axis (24 fruits per treatment in total) then homogenized for the following measurements. Different forms of Ca in the tomato fruits were extracted and determined by adopting Ohat Y's method (Ohat Y, 1970), including Ca nitrate and Ca chloride, water soluble organic Ca, Ca pectate, Ca phosphate and Ca carbonate, Ca oxalate, Ca silicate, hereafter the above Ca forms were recording successively as Alc-Ca, $H_2O$-Ca, NaCl-Ca, HAC-Ca, HCl-Ca, Res-Ca based on the extraction solvent type. Other quality indexes were evaluated by common testing methods (AOAC, 1990; Wang et al., 2011; Yang et al., 2012): Vitamin C content was measured by the 2, 6-dichloroindophenol titrimetric method; soluble sugar was measured by the anthrone method; soluble protein was measured by the Coomassie brilliant blue method; nitrate content was measured by the ultraviolet spectrophotometry method. For each treatment, thirty tomato fruits with a red or orange color were collected randomly to determine the basic morphological parameters, containing long diameter (L-diameter) and short diameter (S-diameter). Fruit weight was calculated from the total number and weight of fruits harvested.

# Results

Calcium Solubility in Lab Experiment Preparation of EM-Calcium Mixtures

According to the survey results from the measured $Ca^{2+}$concentration, the Ca solubility was calculated and showed in Table 2. Results indicated that the measured $Ca^{2+}$concentration of T5 and T4 was significantly higher ($P<0.05$) than that of T1, T2, T3. The Ca solubility of T4 reached the peak value of 89.5%, which was 22.17%, 31.00%, 9.90%, and 9.79% higher than that of T1, T2, T3, and T5 respectively. Therefore T4 was selected as the optimum formula applying to the following field experiment on account of high calcium solubility. The eventually diluted T4 solution used for the field experiment was acid, with the pH value of 4.7.

**Table 2:** Theoretical and measured $Ca^{2+}$concentration after 6 days fermentation

[i] - Means within a column followed by the same letter are not significant at $P<0.05$ (Duncan's multiple range test)

| Treatment | **Theoretical $Ca^{2+}$concentration** | **Measured $Ca^{2+}$concentration** | **Solubility** |
|---|---|---|---|
| T1 | 3.0‰ | 2.02‰d | 67.33% |
| T2 | 4.0‰ | 2.34‰c | 58.50% |
| T3 | 5.0‰ | 3.98‰b | 79.60% |
| T4 | 6.0‰ | 5.37‰ab | 89.50% |
| T5 | 7.0‰ | 5.58‰a | 79.71% |

## *Forms of CA in Fruit*

The Ca accumulation in tomato fruits can directly reflect the effect of EM-calcium solution. Fig. 1gave the content of different forms of Ca in tomato fruits with different treatments. According to the survey results, the EM-calcium application was mainly beneficial for the accumulation of Alc-Ca, $H_2O$-Ca, NaCl-Ca, and HAC-Ca, and which had no obvious effects on the HCl-Ca and Res-Ca accumulation.

Ca pectate (NaCl-Ca) was the main Ca form presented in the ripe tomato fruits, occupied more than 75% of total Ca content, as the figures shown. The NaCl-Ca content of SR, SF, SL, SO was significantly higher ($P<0.05$) than that of CK, and the increases of SL were the most obvious, closely followed by SO, which implied that SL and SO were conductive to the accumulation of NaCl-Ca. While no significant difference ($P>0.05$) for NaCl-Ca content was observed between ST and CK. Water-soluble calcium The EM-calcium application also had great influences on the accumulation of water-soluble Ca including the Alc-Ca and $H_2O$-Ca in tomato fruits. Based on the results, SL increased the content of Alc-Ca and $H_2O$-Ca most significantly ($P<0.05$) by 2.19 and 0.71 multiples respectively in comparison with CK. Besides, SR, SF, SO and ST increased Alc-Ca content significantly ($P<0.05$), and SR, SF increased $H_2O$-Ca content significantly ($P<0.05$), compared to CK. The $H_2O$-Ca content in the tomato fruits of SO and ST was significantly different ($P<0.05$), but there were no significant differences between that of SO and CK, either ST and CK.

HAC-Ca content of SL and SO was significantly higher ($P<0.05$) than that of CK by 65.68% and 36.58% respectively, and the differences among CK, SR, SF and ST were not significant ($P>0.05$). Dissimilarly, HCl-Ca content in SF, SL, SO was 24.77%, 29.05%, 10.05% lower than that in CK, and the effects of ST and SR on the variation of HCl-Ca content were not obvious. Res-Ca accounted for the smallest share of total Ca, except for ST, no significant differences of which were found among the treatments ($P>0.05$), indicating that EM-calcium application had little effects on the Res-Ca accumulation in tomato fruits.

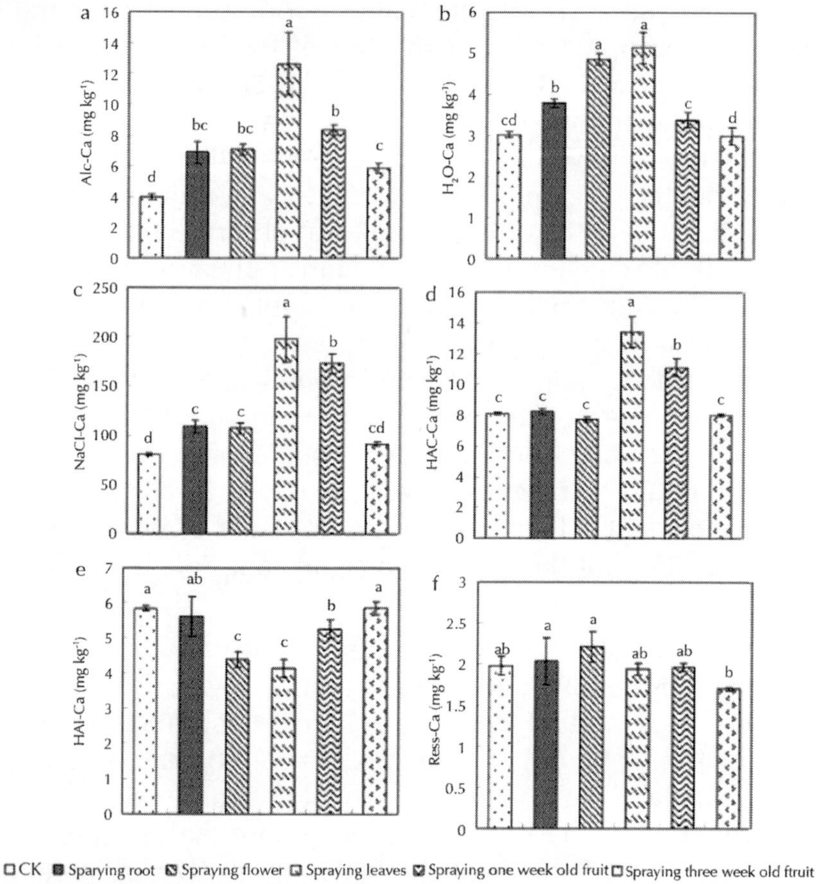


**Figure 1:** The accumulation of different forms of Ca in tomato fruits after the application of EM-calcium solution on different tomato organs. Columns with the same letter represent values that are not significantly different at the 0.05 level of probability according to the Duncan's multiple range test. Each value is the mean ± SD (n=3). The treatment symbols are the same as the experiment design.

## *Fruit Quality*

Fig. 2 showed some quality indexes of tomatoes observed with different spraying methods. Vitamin C (L-ascorbic acid) is essential for all living plants where it functions as the main hydrosoluble antioxidant (Lima-Silva et al., 2012). SF, SL, SO increased the vitamin C of tomato fruit significantly (P<0.05) compared to CK, the increases of SL was most obvious,with the value of 25.38%. However, SR and ST did not affect the content of vitamin C greatly.

Tomato taste quality is largely determined by the contents of soluble sugar (Dorais, 2001). According to the results, the soluble sugar content of SL and SO was significantly higher (P<0.05) than that of CK by 9.65% and 7.20% respectively, but there were no significant differences (P>0.05) among that of CK, SR, SF and ST.

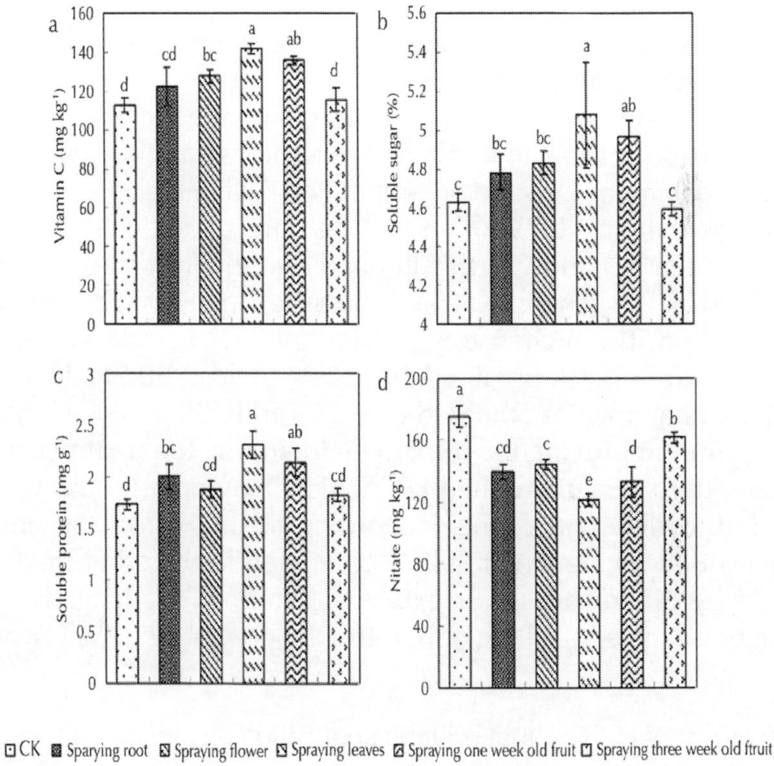


**Figure 2:** Effects of EM-calcium application on some routine indexes of tomato quality. Columns with the same letter represent values that are not significantly different at the 0.05 level of probability according to the Duncan's multiple range test. Each value is the mean ± SD (n=3). The treatment symbols are the same as the experiment design.

SL obtained the highest content of soluble protein, which was significantly higher (P<0.05) than that in CK. SO took the second place, slightly lower than SL but no significant differences ($P>0.05$) were found between them. SF and ST had little effects on the content of soluble protein in tomato fruits. Compared to CK, the nitrate content in the other treatments was significantly lower ($P<0.05$), and SL was particularly apparent. In terms of these basic quality indexes of tomato fruits measured, SL was suggested as the preferable treatment since the above quality indexes of SL were all at the satisfactory levels.

## Fruit Morphology, Yield, and BER Incidence

Fruit appearance is the first quality trait to consumers and determined by fruit size, shape and color (Labate, 2007). The tomato fruit size, yield and BER morbidity harvested from different treatments were displayed in Table 4. Results show that different treatments had little effect on the fruit size, maximum D-value of L-diameter, S-diameter, individual fruit weight was 0.46cm, 0.29cm and 17.8g/fruit respectively, which was found between SO and CK, this implied that the fruit size was mainly determined by the genetic cultivar but has few matters with exogenous Ca application. The highest tomato yield of 72.28 t $ha^{-1}$ was found in SL, followed by SO, CK obtained the lowest yield of 50.26 t $ha^{-1}$, which was 22.02 t $ha^{-1}$ lower compared to SL. Overall, SL processed obvious advantages in improving the tomato yield among the treatments. After applied with EM-Calcium, the BER incidence of tomato fruits was well controlled to different extents, according to the results. SL obtained the lowest BER incidence of 9.42% compared to the other treatments, followed by SR, recording as 10.28%. While ST had relatively less influences on the controlling of BER incidence, only 2.23% lower than CK.

**Table 3:** Effects of EM-Calcium solution application on tomato fruit size, yield and BER incidence

| Treatment | Fruit size | | | Yield (t $ha^{-1}$) | BER incidence (%) |
|---|---|---|---|---|---|
| | L-diameter (cm) | S-diameter (cm) | Weight (g/ fruit) | | |
| CK | 6.18 | 4.66 | 116.70 | 50.26 | 22.81% |
| SR | 6.42 | 4.90 | 128.60 | 56.42 | 10.28% |
| SF | 6.43 | 4.91 | 128.76 | 60.84 | 18.51% |

| | | | | | |
|---|---|---|---|---|---|
| SL | 6.60 | 4.94 | 133.85 | 72.28 | 9.42% |
| SO | 6.64 | 4.95 | 134.50 | 68.09 | 13.22% |
| ST | 6.40 | 4.86 | 126.11 | 53.38 | 20.58% |

## CA Accumulation in Main Parts of Tomato Plant

Ca accumulations in main parts of tomato plant with different EM-Calcium treatments were displayed in Fig. 3; Ca content in upper leaf of SO, ST was significantly higher ($P<0.05$) than that of the other treatments, however, which of SL was slightly lower compared with CK, this indicated that spraying EM-Calcium on tomato leaves may change the migration path of calcium in the leaves; Similar laws were also found in the Ca accumulation of lower leaf, Ca content in lower leaf of SL was significantly lower ($P<0.05$) than that of the other treatments (except CK). According to the results from Ca content in upper leaf and lower leaf, it could be also inferred that SO had the most significant impact on the Ca accumulation of tomato leaves.

ST increased the Ca accumulation of root most significantly, and no significant differences ($P>0.05$) were observed among SF, SL and SO. Ca contents of stem were obviously lower than that of other plant organs, and the differences of which among the treatments were relatively less, Ca content in the stem of SO was significantly higher ($P<0.05$), and there were no significant differences ($P>0.05$) among the other treatments.

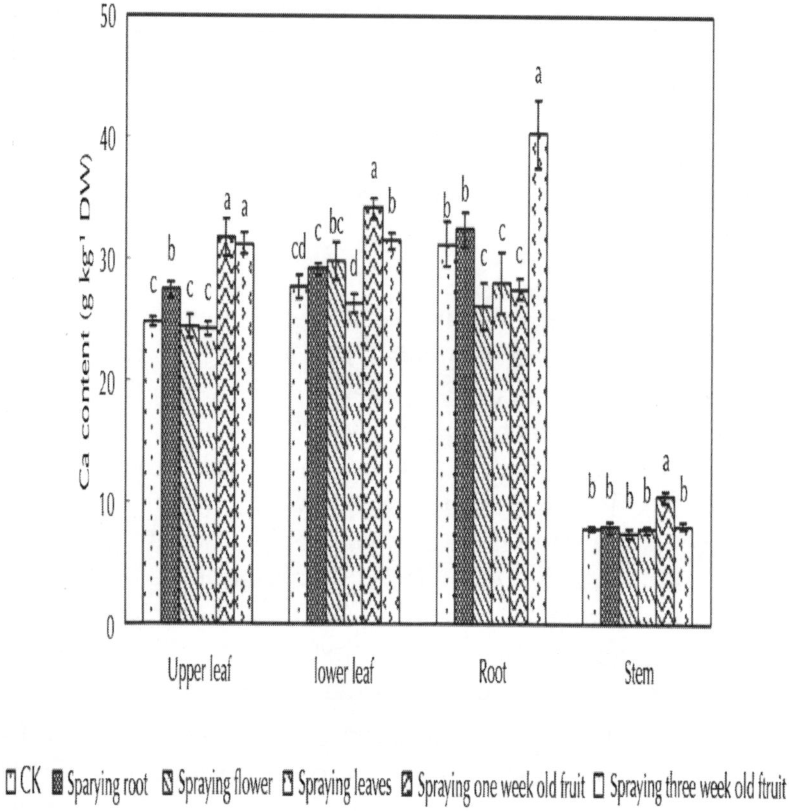


**Figure 3:** Effects of EM-calcium application on the Ca accumulation in main parts of tomato plants. Columns with the same letter represent values that are not significantly different at the 0.05 level of probability according to the Duncan's multiple range test. Each value is the mean ± SD (n=3). The treatment symbols are the same as the experiment design.

# DISCUSSIONS

Calcium is well known for having regulatory roles in metabolism and sodium ions may compete with calcium ions for membranebinding sites (Tuna et al., 2007), and is also known to bind to phospholipids and proteins on the membrane surface, which is required to maintain proper membrane structure and integrity (Jones and Lunt, 1967; Suzuki et al., 2003), thus to some degree, high calcium levels can protect the cell membrane from the adverse effects. In view of the important role calcium played, much emphasis was put on the exogenous Ca application for plants. In the choice of Ca nutrient solution for tomato, materials such as $CaSO_4$ $2H_2O$ (Hall, 1977), $CaNO_3$ $4H_2O$ (Eraslan et al., 2007; Murillo-Amador et al., 2006), $CaCl_2$ (Dong Cai-xia, 2001; Schmitz-Eiberger et al., 2002; Xu et al., 2010), CaO (Almeida et al., 2009; Asiegbu and Uzo, 1983) had been selected as Ca resources for different studies. Before preparing the Ca solution in this study, the activity of Ca ions and cost of the calcium solution were mainly taken into consideration during the selection process of materials, we tried to mixed the gypsum with EM previously, while it was found that the solubility of gypsum in EM was not satisfactory compared to that of lime, the lime was finally selected as the Ca resource. Before the experiment we also concerned about whether the $Ca^{2+}$suspension made by lime would change the survival environment of the acid-loving effective microbes, results proved later that the pH of the mixed liquor fallen back to suitable levels after several days' fermentation. Under microbial actions, the solubility of calcium in EM obtained a satisfactory result (maximum $Ca^{2+}$solubility of 89.50%), while with the increasing application of lime, EM, and molasses, the dissolving capacity of calcium presented a decline trend, it was predicted that the measured $Ca^{2+}$concentration of the EM-Calcium solution would stay in a certain value and the solubility of calcium would decrease with the increasing of raw materials. According to the results of the lab experiment, the mixture of $Ca^{2+}$suspension/ EM/ molasses/ DI water with a volume ratio of 3: 1.5: 1.5: 4 was recommended to practice.

Recent studies tented to adopt foliar-sprayed method on the exogenous Ca application for tomato plants (Eraslan et al., 2007; Gezerel, 1986; Murillo-Amador et al., 2006); there were also experiments (Tabatabaie et al., 2004) about the use of solutions of different concentrations applied to different parts of tomato root system. This experiment showed that the foliar-sprayed method was most beneficial for the Ca accumulation of tomato fruits according to Table 4, spraying EM-Calcium solution on plant leaves increased the Alc-Ca, $H_2O$-Ca, NaCl-Ca, HAC-Ca contents of tomato fruits more significantly than spraying on the other organs, increment of NaCl-Ca content (calcium pectate) was especially notable, this might be related to the increases of soluble pectin in tomato fruits with the ripening of tomatoes (Ashraf M, 1981). Spraying the calcium nutrition on surface of fruits such as litchi, sweet cherry and grape proved not to be an effective way (Combrink, 1995;Huang et al., 2008; Koffmann, 1996), while for tomato fruits in this experiment, results showed that EM-Calcium application had no significant effects on the Ca increment in old tomato fruit, while which had significant effects on the Ca increment in young tomato fruit.

Vitamin C acts as an antioxidant in plants and its levels are responsive to a variety of environmental or stress factors, for example light, temperature, salt and drought, atmospheric pollutants, metals or herbicides (Singh et al., 2012), and which was reported having positive correlation with potassium (K) supply (Marin, 2009), Bangerth (1976) observed an increase in vitamin C content of tomato fruits treated with calcium chloride. In this experiment, we do not exclude the Ca factor when explained the improvement of tomato quality, while EM was more likely to be the main factor, since EM had been reported to have effects on enhancing crop photosynthesis, increasing crop protein contents, and improving crop quality (Daming, 1999;Shousong, 1998). Another evidence to support this speculation was that the soluble sugar correlated negatively with calcium (Beckles, 2012). Taken as a whole, EM-Calcium application significantly improved the fruit quality of tomato (evaluation indexes including vitamin C, soluble sugar, soluble protein, and nitrate); meanwhile, foliar-spray proved to be a preferable method when supplying EM-Calcium in this study.

The BER incidence may induced by the stresses in the root zone, such as salinity, soil water stress, $NH_4^+$toxicity and oxygen withholding (Saure, 2001; Tachibana, 1991), although the impact mechanisms are not fully understood, these factors were considered either directly or indirectly related to $Ca^{2+}$deficiency: Tuna et al.(2007) reported that the exogenous $Ca^{2+}$application significantly improved growth and physiological variables affected by salt stress, from another perspective, Adams (1990) showed that increasing the salinity above 4mS $cm^{-1}$ by addition of major nutrients would reduce Ca content; According to some studies (Albahou, 1999; Žanić et al., 2011), the increased proportions of $NH_4^+$in standard nutrient solution were often associated with severity of blossom-end rot known as a physiological disorder of tomato fruit, and Siddiqi (2002) suggested that the $NH_4^+$presence with a percentage of 10% in total N reduced $Ca^{2+}$accumulation; The BER occured commonly when the soil moisture content was deficit or fully adequate (Adams, 1992, 1993), and there was a minimum rate of transpiration relative to leaf growth rate below which calcium deficiency symptoms were occured (Hamer, 2003); Tachibana (1991) also reported that withholding the oxygen supply to roots at night was a cause of tomato BER, which greatly inhibited the absorption of Ca. The negative correlation between Ca nutrient supply and BER incidence was also found in this study, similar to many other studies (Besford, 1978; Mestre et al., 2012; Olle M, 2009). It was concluded here that the BER incidence of different EM-Calcium treatments was 2.23%-13.39% lower than that of CK.

Fig. 5 showed the yield of marketable tomatoes, which was significantly increased by the treatments except ST, Gezerel (1986) reported the application of foliar fertilizer containing calcium of 0.2% increased the tomato yields and fruit weight. Mayer (2010) inferred that the effects of EM preparation on crop yield increasing could be related to nutrient inputs by EM carrier substrates.Hu and Qi (2013b) reported that long-term effective microorganisms application could promote crop growth and increase yields and nutrition. In this study, we deduced that EM alone had no obvious effects on crop yield, while it could enhance the availability of nutrient supplied. Mestre et al.(2012) observed a negative and significant correlation between fruit yield and BER (r=-0.810), our results showed that the BER incidence had significant correlation with total tomato yield (r=-0.736) and marketable tomato yield (r=-0.862).

Early study (del Amor and Marcelis, 2006) reported that Ca concentration of tomato plant was significantly reduced by low-Ca supply (0.5 meq $L^{-1}$) compared with the nutrient standard solution (9 meq $L^{-1}$), and with 14 days' low-Ca application, Ca concentration in all plant organs (leaves, stems and roots) was reduced by approximately 70% compared to control plants; we reported that 2.0‰ Ca application increased the Ca accumulation in upper leaf, lower leaf, root, stem to maximum rates of 28.09%, 23.50%, 29.15%, 33.34% compared to no-calcium treatment, the causes of the difference were probably related to the nutrient supply methods, calcium supply through roots increased the Ca content of tomato fruits by increasing which of other organs, while calcium spray on leaves or young fruits increased Ca content of tomato fruits by changing the Ca migration, the evidence to support this speculation in this study was that the foliar spray increased the Ca content of fruits but decreased which of leaves. Dong (2001) guessed that a "Ca-attracted" center was formed in the spray organ when spraying Ca nutrient ($CaCl_2$), $Ca^{2+}$was attracted to the center and then migrated from the center to the organ which needed Ca most. However, the migration mechanism about Ca migration with exogenous Ca application needed to be examined by $^{45}Ca$ tracing technique (Behling et al., 1989; Yamauchi et al., 1986).

# EFFECTS OF EM-CALCIUM ON PRODUCTION OF FLUE-CURED TOBACCO

## Materials and Methods

### *Test Site*

The experiments were carried out in a plastic sheet covered greenhouse from March 2013 to September 2013 in the Vegetables and Flowers Institute of Nanjing (latitude 31°43′ N, longitude 118°46′E), China. The average annual rainfall is about 1106.5mm, with the rainy season from the end of June to the middle of July, and the average yearly temperature is approximately 15.7°C and average humidity is about 81%. The soil characters were: pH 5.68, 14.47 g/kg organic matter, 28.5 mg/kg available P, 153.84 mg/kg available K, 1.3 8g/kg total nitrogen.

### *Experimental Design*

K326 was chosen as the flue-cured tobacco plant material, with young plants elaborately cultivated in made-in-order seedling trays; then they would be transplanted into the lysimeters when they grew 6 expanded leaves. The planting density was 12 plants per treatment, with the line spacing of 0.8m and plant spacing of 0.6m. After that, conventional field management was conducted in the first week.

For simulating the water stress in growing stages, the irrigation amount was designed as 400mm during the whole growth stage.

Irrigation waters of root-extending stage, vigorous stage and maturity stage in this experiment were assigned as 40%, 20%, 40% for the total water consumption respectively, and they were irrigated 6 times in each growth stage. Tobacco dedicated fertilizers (provided by the Institute of Guizhou Tobacco Science, N: $P_2O_5$: $K_2O$=1:2:3) were applied according to a proportion of basal dressing: topdressing=7:3, the latency time of topdressing was 26 days after transplanted.

Detailed experimental treatments regarding the water-retaining agent amount and EM-calcium content were shown in Table 4. EM-calcium with 2.0‰ $Ca^{2+}$concentration was prepared as 2.2.1.

**Table 4:** Experimental design

| Treatment | **$Ca^{2+}$concentration** | **EM-Ca amount (ml/time)** | **Intervals of spraying (days)** | **MP3005 (g/plant)** | **Growth stage** |
|---|---|---|---|---|---|
| CK | --- | --- | --- | --- | --- |
| P1 | 1‰ | 2 | 3 | 0 | Root extending (R) |
| P2 | 1‰ | 2 | 3 | 0 | Vigorous (V) |
| P3 | 1‰ | 2 | 3 | 0 | R+V |
| P4 | 1‰ | 2 | 3 | 30 | R |
| P5 | 1‰ | 2 | 3 | 30 | V |
| P6 | 1‰ | 2 | 3 | 30 | R+V |

## Measurements

3 representative tobacco plants were sampled for each replication. At harvest time, the lower leaves, middle leaves and upper leaves were collected orderly, killed by 105°C high temperature and toasted to the constant weight (Hou et al., 2012; Maomao Hou, 2013). Weight of dry tobacco leaves was measured and recorded to calculate the total yield.

*IWUE* was calculated by the formula (Aujla et al., 2005; Ünlü et al., 2011):

$$IWUE = \frac{Y}{I}$$

Where, *IWUE* ($kg/m^3$) was the irrigation water use efficiency; *Y* was the tobacco dry yield ($kg/hm^2$); I was the irrigation amount ($m^3/hm^2$).

# Results

## Agronomic Characters of Flue-Cured Tobacco

Fig. 4 showed the changes of total area of flue-cured tobacco leaves in single plant with days after transplanted. As shown in the Figure 4, during 45~77 days, flue-cured tobaccos in P4 grew more satisfactory compared to that in other treatments, and the leaf area of single plant in T4 reached a higher value of 36897.9 $cm^2$ in 77 days. However, P2 showed a poor performance in leaf area enhancement, recording as 32110.1 $cm^2$ only. From the view of leaf area enhancement, the effects of water-retaining agent treatments were obviously better than those with no water-retaining agent. Under the same application amount of water-retaining agent, the effects of EM-calcium spray in root-extending stage were better than those in vigorous stage and root-extending+vigorous stage. The results were different from the expectation, it was maybe that the EM-calcium spray blocked the leaf stoma after long-time application. On the whole, P4 was supposed to be the better treatment from a pure view of leaf area increasing.

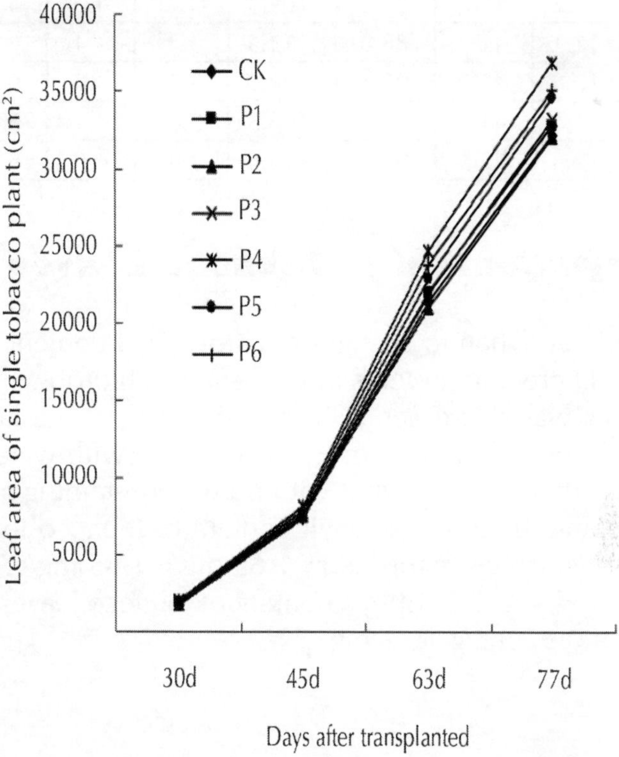


**Figure 4:** Leaf area increasing with days after transplanted.

Table 5 displayed the plant height and stem girth with days after transplanted under different treatments. As shown from the Table 5, no obvious differences were found among the treatments, this was mainly because the fertilizer application was equal in different treatments. At 77 days after transplanted, the plant height of different treatments were recorded as 105.4~ 122.0 cm and the plant stem girth were around 6.9~7.4 cm.

**Table 5:** Plant stem girth and height with days after transplanted

| Treatment | **Height** (cm) | | | | **Stem** (cm) | | | |
|---|---|---|---|---|---|---|---|---|
| | **30**d | **45**d | **63**d | **77**d | **30**d | **45**d | **63**d | **77**d |
| CK | 9.2 | 38.7 | 84.5 | 109.9 | 3.1 | 4.3 | 6.0 | 7.0 |
| P1 | 10.5 | 39.6 | 86.2 | 110.6 | 3.2 | 4.3 | 6.3 | 7.2 |

| | | | | | | | | |
|---|---|---|---|---|---|---|---|---|
| P2 | 10.2 | 35.8 | 83.9 | 105.4 | 3.1 | 4.0 | 5.9 | 6.9 |
| P3 | 11.1 | 41.2 | 88.0 | 113.9 | 3.3 | 4.4 | 6.2 | 7.3 |
| P4 | 12.8 | 43.9 | 97.4 | 122.0 | 3.7 | 4.8 | 6.5 | 7.4 |
| P5 | 11.4 | 41.2 | 91.8 | 108.5 | 3.4 | 4.5 | 6.3 | 7.1 |
| P6 | 12.3 | 42.5 | 96.9 | 119.8 | 3.4 | 4.6 | 6.6 | 7.2 |

## Chlorophyll Content of Tobacco Leaves

Fig. 5 showed the dynamic changes of chlorophyll content in tobacco leaves under different treatments more visually, chlorophyll content in tobacco leaves was decreased with a higher rate at the later growth stage. The chlorophyll content of tobacco leaves with water-retaining agent was slightly lower than that with no water-retaining agent, at 63 d after transplanted, the chlorophyll content of tobacco leaves under different treatments was recorded as 1.69 mg/g-1.85 mg/g; at 87 days after transplanted, the chlorophyll content of tobacco leaves was in the lowest level of 0.97 mg/g-1.12 mg/g.

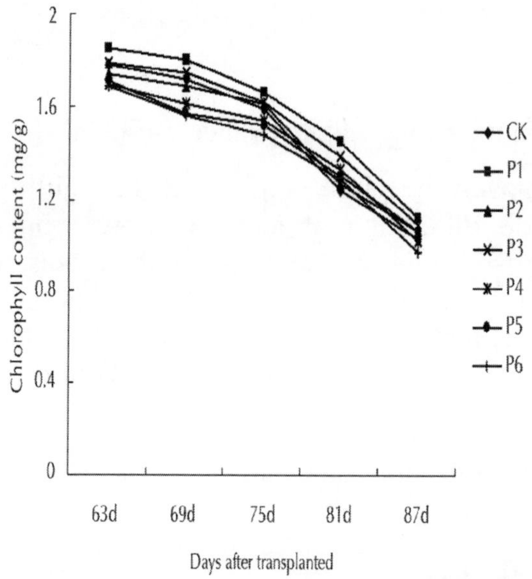


**Figure 5:** The changes of chlorophyll content in tobacco leaves varying with days after transplanted.

## Dry Matter Accumulation

Table 6 showed the dry matter accumulation of different tobacco organs and the total yield under different treatments. From the table it was found that the dry matter amount of leaf and stem and root in P4 was the highest, recording as 73.6 g, 88.1 g and 164.9 g respectively. The dry matter amount of leaves with water-retaining agent treatments were 4.99%-14.28% higher than that with no water-retaining agent application. Under the same amount of water--retaining agent application, EM-calcium spray in root-extending stage was much better than that in other stages. Dry matter amount of leaf in P1 and P4 was higher than that of other treatment, and the tobacco yield of P4 was highest, recording as 2473.5 kg/hm$^2$, followed closely by P6, which of P2 was the lowest.

**Table 6:** The dry matter accumulation of different tobacco organs and the total yield under different treatments

| Treatment | **Root** | | **Stem** | | **Leaf** | | **Whole Plant(g)** | **Yield (kg/ hm$^2$)** |
|---|---|---|---|---|---|---|---|---|
| | **Dry matter(g)** | **Proportion (%)** | **Dry matter (g)** | **Proportion (%)** | **Dry matter (g)** | **Proportion(%)** | | |
| CK | 63.2 | 21.76% | 82.9 | 28.55% | 144.3 | 49.69% | 290.4 | 2164.5 |
| P1 | 62.1 | 21.01% | 85.4 | 28.89% | 148.1 | 50.10% | 295.6 | 2221.5 |
| P2 | 58.7 | 21.43% | 75.4 | 27.53% | 139.8 | 51.04% | 273.9 | 2097.0 |
| P3 | 64.5 | 22.13% | 80.5 | 27.62% | 146.5 | 50.26% | 291.5 | 2197.5 |
| P4 | 73.6 | 22.54% | 88.1 | 26.97% | 164.9 | 50.49% | 326.6 | 2473.5 |
| P5 | 67.7 | 22.76% | 78.3 | 26.32% | 151.5 | 50.92% | 297.5 | 2272.5 |
| P6 | 68.7 | 22.50% | 77.9 | 25.52% | 158.7 | 51.98% | 305.3 | 2380.5 |

## Nutrient Absorption

Calcium is well known to maintain proper membrane structure and integrity and plays important roles in crop development and disease control (Almeida et al., 2009; Berry et al., 1988; Evans, 1953; Hall, 1977). In flue-cured tobacco cultivation, calcium helps to coordinate the physiology function of tobacco plant, making the tobacco plant root system stronger, growing vigorously and harvesting timely. $Ca^{2+}$has promoting effects for the growth of tobacco seedlings and can improve the drought-resistant ability of tobacco seedlings. In addition, the high calcium content in tobacco leaves delays the maturity of tobacco, characterized by stiffness and hardness of the tobacco leaves, thus the use value of tobacco leaves is decreased, excess calcium may also cause disorder of some microelements in tobacco plants and produce toxic impacts; while calcium deficiency can lead to the deformity of tobacco plants and generate a spoon-shaped reverse disease of tobacco leaves. As was shown in Fig. 6, the exogenous application of calcium significantly increased the calcium content in tobacco leaves, and the effects of spraying in root-extending stage were more satisfactory than those in other stages, this may be related to that the water stress in maturity affected negatively the effects of EM-calcium.

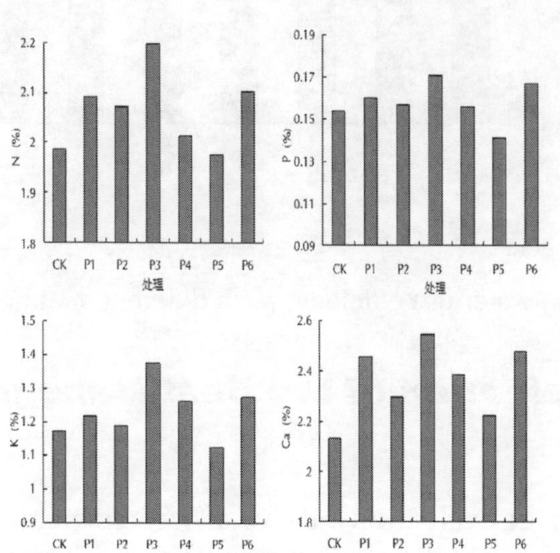


**Figure 6:** Nutrient absorption of tobacco leaves with different treatments.

## *Irrigation Water Use Efficiency*

Fig. 7 showed the *IWUE* of flue-cured tobacco plants with different treatments. It could be seen that the differences of *IWUE* among the treatments were more significant, P4 obtained the highest *IWUE* of 0.618 kg/m$^3$, followed by P6, and *IWUE* value of T2 was the minimum, recording as 0.524 kg/m$^3$. Since the equal irrigation amount among the treatments, *IWUE* presented a positive relationship with the flue-cured tobacco yield.

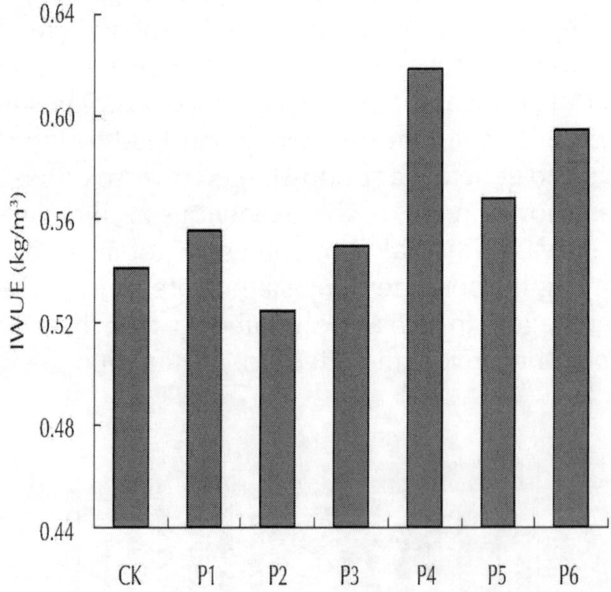


**Figure 7.**Irrigation water use efficiency with different treatments.

## *Optimal Selection of the Best Management Scheme*

Table. 7 showed the evaluation indexes of different treatments including the tobacco yield, *IWUE,*Ca content in tobacco leaves and the cost, among these indexes, the cost was calculated based on the price list of raw materials provided by EMRO Limited Company, Nanjing

Branch. Table. 7showed that the advantages and disadvantages of each treatment were distinct, taking P1, P2 and P3 as example, although the cost of them was lower, the yield and *IWUE* were lower than those of P4, P5 and P6; Ca content of tobacco leaves with P3 treatment was the highest, reaching 2.546%, but the yield of P3 was not in the highest level. Therefore, a scientific evaluation method was needed here to select the optimal management scheme.

**Table 7:** Evaluation indexes

| Treatment | **Yield (kg/hm²)** | **IWUE (kg/m³)** | **Ca (%)** | **Cost (USD/plant)** |
|---|---|---|---|---|
| CK | 2164.5 | 0.54 | 2.131 | 0 |
| P1 | 2221.5 | 0.56 | 2.454 | 0.0033 |
| P2 | 2097.0 | 0.52 | 2.297 | 0.0033 |
| P3 | 2197.5 | 0.55 | 2.546 | 0.0067 |
| P4 | 2473.5 | 0.62 | 2.387 | 0.0167 |
| P5 | 2272.5 | 0.57 | 2.224 | 0.0167 |
| P6 | 2380.5 | 0.60 | 2.478 | 0.0200 |

Modeling approach was shown below (Chen and Li, 2010; Chou et al., 2012):

Supposing that there were *n* evaluation indexes and *m* schemes, *m* schemes corresponding with *n*indexes obtained the following matrix: $R=(r_{ij})_{m\times n}$

Where; $r_{ij}$ is the $j^{th}$ evaluation index of the $i^{th}$ scheme. To $r_{j}$, there was information entropy:

$$E_j = -\sum_{i=1}^{m} p_{ij} \ln p_{ij}, (j = 1,2,3,\ldots n)$$

And $P_{ij}$ were calculated as the formula:

$$p_{ij} = r_{ij} / \sum_{i=1}^{m} r_{ij}$$

The entropy value of $j^{th}$ index was:

$$e_j = \frac{1}{\ln m} Ej, (j=1,2,3,\ldots,n)$$

The objective weight of $j^{th}$ index was:

$$\theta j = (1-e_j) / \sum_{i=1}^{n} (1-e_j), (j=1,2,3,\ldots,n)$$

It was clear that:

$$0 \le \theta j \le 1; \sum_{j=1}^{n} \theta_j = 1$$

This study took the subjective information into the calculations, the comprehensive weight could be obtained by combining the subjective weight *w1, w2, w3…wn* of the decision makers with the objective weight $\theta j$ *(j=1,2,3,…,n)*:

$$\alpha_j = \theta_j \bar{w}_j \mid \sum_{j=1}^{n} \theta_j \bar{w}_j, (j = 1,2,3,\ldots,n)$$

Recording the optimum value of each row as $rj^*$, normalize the elements in the matrix, and $rj^*$ value was varied with the index characters. The indexes could be divided into two classes: The profitable indexes and the damnous indexes, which were "the larger the better" and "the smaller the better", listing as follows:

$$d_{ij} = \begin{cases} \dfrac{r_{ij}}{r_j *}, r_j * = \max\{r_{ij}\} \\ \dfrac{r_j *}{r_{ij}}, r_j * = \min\{r_{ij}\} \end{cases}$$

The entropy coefficient value (A better management scheme would obtain a higher entropy coefficient value) of each treatment could be calculated by:

$$\lambda_i = \sum_{j=i}^{n} \alpha d_{ij}, i = 1,2,3,\ldots,m.$$

The calculated objective weight of the tobacco yield, *IWUE,* Ca content in tobacco leaves and the cost were 0.2142, 0.2135, 0.2003, and 0.3720. However, since the strict requirement on water saving, the subjective weight of which was assigned as 0.4, 0.4, 0.15, 0.05 respectively. Fig. 8 showed the entropy weight coefficient of different treatments, based on the principle of entropy weight coefficient "the higher the better", P4 was supposed to be the best scheme, in other words, 30 g/plant MP3005 water-retaining agent combined with EM-calcium spray (‰1 $Ca^{2+}$) during the root-extending stage of flue-cured tobaccos was the optimal management scheme, and the interval of spraying time was 3 days with 2 mm each time on the back side of tobacco leaves. Additionally, entropy weight coefficient value of P1 and P4 were similar, but the mechanism was different, P1 tended to obtain a lower cost, and P4 tended to obtain a higher yield and *IWUE.*

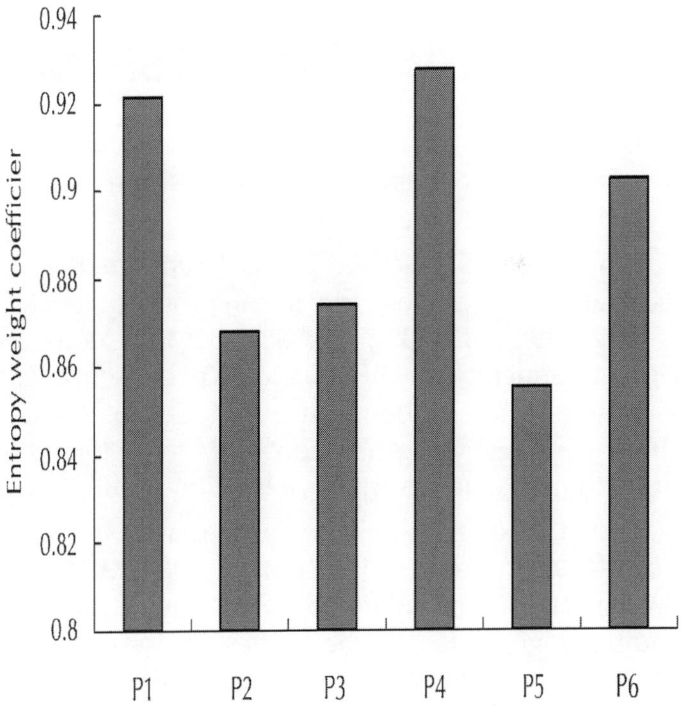


**Figure 8:** Entropy weight coefficient of different treatments.

# CONCLUSIONS

Treatment P4 (spraying 1‰ EM-Ca amount 2 ml with 30 g MP3005 per plant once in 3 days during tobacco root extending stage) obtained the highest flue-cured tobacco yield of 2473.5 $kg/hm^2$, followed by P6, and yield of P2 was lowest, recording as only 2097.0 $kg/hm^2$. The irrigation water use efficiency of P4 was highest, reaching 0.618 $kg/m^3$. Exogenous Ca supply significantly increased the Ca content in tobacco leaves. The evaluation results of entropy weight coefficient evaluation model showed that P4 was the best management scheme, that was to say, 30 g/plant MP3005 water-retaining agent combined with EM-calcium spray (1‰ $Ca^{2+}$) during the root-extending stage of flue-cured tobaccos was the optimal management scheme, and the interval of spraying time was 3 days with 2ml each time on the back side of tobacco leaves.

# GENERAL CONCLUSIONS

The research results showed that Active EM-Calcium could promote crop growth, improve the yield and disease resistance of crops. Main conclusions could be drawn as below:

- The application of Active EM-Calcium increased the Ca accumulation in upper leaf, lower leaf, root and stem to maximum rates of 28.09%, 23.50%, 29.15%, 33.34% compared to no-calcium treatment. The BER incidence of Active EM-Calcium treatments was lower than that of CK. The BER incidence had significantly negative correlation with total tomato yield (r=-0.736) and marketable tomato yield (with no BER) (r=-0.862).
- Exogenous Ca supply significantly increased the Ca content in tobacco leaves. Treatment P4 (spraying 1‰ EM-Ca amount 2 ml with 30 g MP3005 per plant once in 3 days during tobacco root extending stage) obtained the highest flue-cured tobacco yield of 2473.5 $kg/hm^2$, and the highest irrigation water use efficiency with 0.618 $kg/m^3$.

# ACKNOWLEDGEMENTS

This work was financed by the Fund of the Ministry of Water Resources Public Welfare Project (201301017), the Fundamental Research Funds for the Central Universities (2013/B13020057), the Graduate Student Research and Innovation Program of Jiangsu Province (257) and Creative Funds of Jiangsu province (CXZZ13_0266).


# REFERENCES


1. Adams, P., 1990. Effect of salinity on the distribution of calcium in tomato (Lycopersicon esculentum) fruit and leaves, In: Beusichem, M.L. (Ed.), Plant Nutrition — Physiology and Applications. Springer Netherlands, . 473-476.
2. Adams, P., Ho, L.C., 1992. The susceptibility of modern tomato cultivars to blossom-end rot in relation to salinity. J. Hor. Sci. 67, 827-839.
3. Adams, P., Ho, L.C., 1993. Effects of environment on the uptake and distribution of calcium in tomato and on the incidence of blossom-end rot. Plant Soil. 154, 127-132.
4. Albahou, M., 1999. Alternative greenhouse tomato production system. ph. D. Thesis. Oregon State University. Oregon, 223
5. Almeida, G.D., Pratissoli, D., Zanuncio, J.C., Vicentini, V.B., Holtz, A.M., Serrão, J.E., 2009. Calcium silicate and organic mineral fertilizer increase the resistance of tomato plants to Frankliniella schultzei. Phytoparasitica. 37, 225-230.
6. AOAC, 1990. Official Methods of Analysis, 15th ed. Association of Official Analytical Chemists. Washington, DC.
7. Ashraf, M. K.N., 1981. Studies on the pectinesterase activity and some chemical constituents of some Pakistan mango varieties during storage ripening. J. Agric. Food Chem. 29, 526-528.
8. Asiegbu, J.E., Uzo, J.O., 1983. Effects of lime and magnesium on tomato (Lycopersicon esculentum Mill) grown in a ferrallitic sandy loam tropical soil. Plant Soil. 74, 53-60.
9. Aujla, M.S., Thind, H.S., Buttar, G.S., (2005). Cotton yield and water use efficiency at various levels of water and N through

drip irrigation under two methods of planting. Agricultural Water Management. 71, 167-179.

10. Bangerth, F., 1976. Relationship between calcium content and the content of ascorbic acid in apple, pear and tomato fruits. Qual. Plant. 26, 341-348.
11. Besford, R.T., 1978. Effect of potassium nutrition of three tomato varieties on incidence of blossom-end rot. Plant Soil. 50, 179-191.
12. Berry, S.Z., Madumadu, G.G., Uddin, M.R., 1988. Effect of calcium and nitrogen nutrition on bacterial canker disease of tomato. Plant Soil. 112, 113-120.
13. Behling, J., Gabelman, W.H., Gerloff, G.C., 1989. The distribution and utilization of calcium by two tomato (Lycopersicon esculentum Mill.) lines differing in calcium efficiency when grown under low-Ca stress. Plant Soil. 113, 189-196.
14. Bennett, A.J., Mead, A., Whipps, J.M., 2009. Performance of carrot and onion seed primed with beneficial microorganisms in glasshouse and field trials. Biol Control. 51, 417-426.
15. Beckles, D.M., 2012. Factors affecting the postharvest soluble solids and sugar content of tomato (Solanum lycopersicum L.) fruit. Postharvest Biol Tec. 63, 129-140.
16. Berry, S.Z., Madumadu, G.G., Uddin, M.R., (1988). Effect of calcium and nitrogen nutrition on bacterial canker disease of tomato. Plant Soil. 112, 113-120.
17. Chen, T.Y., Li, C.H., 2010. Determining objective weights with intuitionistic fuzzy entropy measures: A comparative analysis. Information Sciences. 180, 4207-4222.
18. Chou, Y.C., Yen, H.Y., Sun, C.C., 2012. An integrate method for performance of women in science and technology based on entropy measure for objective weighting. Qual Quant, 1-16.
19. Chiasson, D., Ekengren, S., Martin, G., Dobney, S., Snedden, W., 2005. Calmodulin-like Proteins from Arabidopsis and Tomato are Involved in Host Defense Against Pseudomonas syringae pv. tomato. Plant Mol Biol. 58, 887-897.
20. Chung, M., Han, J.S., Giovannoni, J., Liu, Y., Kim, C., Lim, K., Chung, J., 2010. Modest calcium increase in tomatoes expressing

a variant of Arabidopsis cation/H+ antiporter. Plant Biotechnol Rep. 4, 15-21.

21. Combrink, N.J.J., Jacobs, G., Maree, P.C.J., 1995. The effect of calcium and boron on the quality of muskmelons (Cucumis melo L.). Journal of Southern African Society for Horticultural Science. 5, 33-38.
22. Daly, M.J., Stewart, D.P.C., 1999. Influence of "Effective Microorganisms" (EM) on Vegetable Production and Carbon Mineralization–A Preliminary Investigation. J. Sustain Agr. 14, 15-25.
23. Daming, R., 1999. Studies on the action mechanism of effective microorganisms in crop production. Journal of maize science in China. 7, 62-64.
24. Del Amor, F.M., Marcelis, L.F.M., 2006. Differential effect of transpiration and Ca supply on growth and Ca concentration of tomato plants. Sci Hortic. 111, 17-23.
25. Dong C.X, Z.J.M., Fan X.H., Wang H.Y. 2001. Effects of different ways of Ca supplements on the Ca content and forms in mature fruits of tomato. Plant Nutrition and Fertilizer Science. 10, 91-95.
26. Dorais, M., Papadopoulos, A., Gosselin, A., 2001. Greenhouse tomato fruit quality. Hort Reviews. 26, 239-319.
27. Eraslan, F., Akbas, B., Inal, A., Tarakcioglu, C., 2007. Effects of foliar sprayed calcium sources onTomato mosaic virus (ToMV) infection in tomato plants grown in greenhouses. Phytoparasitica. 35, 150-158.
28. Evans, H.J., 1953. Relation of calcium nutrition to the incidence of blossom-end rot in tomatoes. proc. Am. Soc. Hor. Sci. 61, 346-352.
29. Freitas, S., Jiang, C.Z., Mitcham, E., 2012. Mechanisms Involved in Calcium Deficiency Development in Tomato Fruit in Response to Gibberellins. J Plant Growth Regul. 31, 221-234.
30. Gezerel, Ö., 1986. The Effect of Calcium-Containing Foliar Fertilizers on Tomato Yields, In: Alexander, A. (Ed.), Foliar Fertilization. Springer Netherlands, . 304-309.
31. Hall, D.A., 1977. Some effects of varied calcium nutrition on the growth and composition of tomato plants. Plant Soil. 48, 199-211.

32. Hamer, P.J.C., 2003. Analysis of strategies for reducing calcium deficiencies in glasshouse grown tomatoes: model functions and simulations. Agr Syst. 76, 181-205.
33. Heo, S.-U., Moon, S.-Y., Yoon, K.-s., Kim, Y.-J., Koo, Y.-M., 2008. Enhanced compost maturity by effective microorganisms. J Biotechnol. 136, Supplement, S65.
34. Higa, T., 1997. Great revolution to save the earth in Chinese edition. China Agriculture Publisher University.
35. Hou, M.M., Shao, X.H., Chen, L.H., Chang, T.T., Wang, W.N., Yang, Q., Wang, Y.F., 2012. Study on nitrogen utilization efficiency of flue-cured tobacco with 15n tracing technique, 1st International Conference on Energy and Environmental Protection, 520-524.
36. Hou, M.M., Shao X.H., Zhai Y.M. 2013. Entropy weight coefficient evaluation of comprehensive index for flue-cured tobacco and its response to different water-nitrogen treatments. Research on Crops. 14, 1232-1237.
37. Huang, X.M., Wang, H.C., Zhong, W.L., Yuan, W.-Q., Lu, J.M., Li, J.G., 2008. Spraying calcium is not an effective way to increase structural calcium in litchi pericarp. Sci Hortic. 117, 39-44.
38. Hu, C., Qi, Y., 2013a. Effective microorganisms and compost favor nematodes in wheat crops. Agron. Sustain. Dev., 1-7.
39. Hu, C., Qi, Y., 2013b. Long-term effective microorganisms application promote growth and increase yields and nutrition of wheat in China. European Journal of Agronomy. 46, 63-67.
40. Jones, R.G.W., Lunt, O.R., 1967. The function of calcium in plants. Bot. Rev. 33, 407-426.
41. Javaid, A., 2010. Beneficial Microorganisms for Sustainable Agriculture, In: Lichtfouse, E. (Ed.), Genetic Engineering, Biofertilisation, Soil Quality and Organic Farming. Springer Netherlands, pp. 347-369.
42. Khaliq, A., Abbasi, M.K., Hussain, T., 2006. Effects of integrated use of organic and inorganic nutrient sources with effective microorganisms (EM) on seed cotton yield in Pakistan. Bioresource Technology. 97, 967-972.
43. Koffmann, W., Wade, N.L., Nicol, H., 1996. Tree sprays and root pruning fail to control rain induced cracking of sweet cherries. Plant Protection Quarterly. 11, 126-130.

44. Kumar, S., Dey, P., 2011. Effects of different mulches and irrigation methods on root growth, nutrient uptake, water-use efficiency and yield of strawberry. Scientia Horticulturae. 127, 318-324.

45. Labate, J.A., Grandillo, S., Fulton, T., Mun´ os, S., Caicedo, A., Peralta, I., et al., 2007. Tomato. In: Kole, C. (Ed.),. Genome Mapping and Molecular Breeding in Plants. vol. 5. Springer, New York, pp. , 1–125.

46. Lima-Silva, V., Rosado, A., Amorim-Silva, V., Muñoz-Mérida, A., Pons, C., Bombarely, A., Trelles, O., Fernández-Muñoz, R., Granell, A., Valpuesta, V., Botella, M., 2012. Genetic and genome-wide transcriptomic analyses identify co-regulation of oxidative response and hormone transcript abundance with vitamin C content in tomato fruit. BMC Genomics. 13, 1-15.

47. Marin, A., Rubio, J. S., Martinez, V., & Gil, M. I., 2009. Antioxidant compounds in green and red peppers as affected by irrigation frequency, salinity and nutrient solution composition. J Sci Food Agr. 89, 1352-1359.

48. Mayer, J., Scheid, S., Widmer, F., Fließbach, A., Oberholzer, H.R., 2010. How effective are 'Effective microorganisms® (EM)'? Results from a field study in temperate climate. Appl Soil Ecol. 46, 230-239.

49. Mestre, T.C., Garcia-Sanchez, F., Rubio, F., Martinez, V., Rivero, R.M., 2012. Glutathione homeostasis as an important and novel factor controlling blossom-end rot development in calcium-deficient tomato fruits. J Plant Physiol. 169, 1719-1727.

50. Mowa, E., Maass, E., 2012. The effect of sulphuric acid and effective micro-organisms on the seed germination of Harpagophytum procumbens (devil›s claw). S Afr J Bot. 83, 193-199.

51. Murillo-Amador, B., Jones, H.G., Kaya, C., Aguilar, R.L., García-Hernández, J.L., Troyo-Diéguez, E., Ávila-Serrano, N.Y., Rueda-Puente, E., 2006. Effects of foliar application of calcium nitrate on growth and physiological attributes of cowpea (Vigna unguiculata L. Walp.) grown under salt stress. Environ Exp Bot. 58, 188-196.

52. Ohat Y, Y.K., Deguchi M, 1970. Chemical fractionation of calcium in the fresh leaf blade and influences of deficiency or over supply of calcium and age of leaf on the content of each calcium fraction. J. Sci. Soil Manure. 41, 19-26.

53. Olle M, B.I., 2009. Causes and control of calcium deficiency disorders in vegetables: a review. J Hort Sci Biotechnol. 84, 577-584.
54. Raleigh, S.M., Chucka, J.A., 1944. Effect of nutrient ratio and concentration on growth and composition of tomato plants and on the occurrence of blossom-end rot of the fruit. Plant Physiol. 19, 671-678.
55. Sachan, R.S., Sharma, R.B., 1981. An easy estimation of tomato root parameters based on calcium absorption. Biol Plant. 23, 311-314.
56. Shao X.H., H.M.M., .Chen .J.N. 2013. Effects of EM-calcium spray on Ca uptake, blossom-end rot incidence and yield of greenhouse tomatoes (Lycopersicon esculentum). Research on Crops. 14, 1159-1166.
57. Shousong, Y., 1998. The effect of effective microorganisms spray on soybean yield and quality. Liaoning agricultural science in China. 3, 12-13.
58. Saure, M.C., 2001. Blossom-end rot of tomato (Lycopersicon esculentum Mill.) — a calcium- or a stress-related disorder? Sci Hortic. 90, 193-208.
59. Schmitz-Eiberger, M., Haefs, R., Noga, G., 2002. Calcium deficiency - Influence on the antioxidative defense system in tomato plants. J Plant Physiol. 159, 733-742.
60. Siddiqi, M.Y., Malhotra, B., Min, X., Glass, A.D.M, 2002. Effects of ammonium and inorganic carbon enrichment on growth and yield of a hydroponic tomato crop. Plant Nutr. Soil Sci. 165, 191-197.
61. Suzuki, K., Shono, M., Egawa, Y., 2003. Localization of calcium in the pericarp cells of tomato fruits during the development of blossom-end rot. Protoplasma. 222, 149-156.
62. Sonneveld, C., Voogt, W., 2009. Calcium Nutrition and Climatic Conditions, Plant Nutrition of Greenhouse Crops. Springer Netherlands, 173-201.
63. Singh, D.P., Beloy, J., McInerney, J.K., Day, L., 2012. Impact of boron, calcium and genetic factors on vitamin C, carotenoids, phenolic acids, anthocyanins and antioxidant capacity of carrots (Daucus carota). Food Chem. 132, 1161-1170.

64. Tachibana, S., 1991. Import of calcium by tomato fruit in relation to the day-night periodicity. Sci Hortic. 45, 235-243.
65. Tabatabaie, S.J., Gregory, P.J., Hadley, P., 2004. Uneven distribution of nutrients in the root zone affects the incidence of blossom end rot and concentration of calcium and potassium in fruits of tomato. Plant Soil. 258, 169-178.
66. Taylor, M.D., Locascio, S.J., 2004. Blossom-end rot: a calcium deficiency. J. Plant Nutr. 27, 123-139.
67. Truax, B., Gagnon, D., 1993. Effects of straw and black plastic mulching on the initial growth and nutrition of butternut, white ash and bur oak. Forest Ecology and Management. 57, 17-27.
68. Tuna, A.L., Kaya, C., Ashraf, M., Altunlu, H., Yokas, I., Yagmur, B., 2007. The effects of calcium sulphate on growth, membrane stability and nutrient uptake of tomato plants grown under salt stress. Environ Exp Bot. 59, 173-178.
69. Ünlü, M., Kanber, R., Koç, D.L., Tekin, S., Kapur, B., 2011. Effects of deficit irrigation on the yield and yield components of drip irrigated cotton in a mediterranean environment. Agricultural Water Management. 98, 597-605.
70. Wang, F., Kang, S., Du, T., Li, F., Qiu, R., 2011. Determination of comprehensive quality index for tomato and its response to different irrigation treatments. Agr Water Manage. 98, 1228-1238.
71. Xu, T., Li, T., Qi, M., 2010. Calcium effects on mediating polygalacturonan activity by mRNA expression and protein accumulation during tomato pedicel explant abscission. Plant Growth Regul. 60, 255-263.
72. Yamauchi, T., Hara, T., Sonoda, Y., 1986. Effects of boron deficiency and calcium supply on the calcium metabolism in tomato plant. Plant Soil. 93, 223-230.
73. Yamazaki, H. H., T., 1995. Calcium nutrition affects resistance of tomato seedlings to bacterial wilt. HortScience. 30, 91-93.
74. Yang, L., Qu, H., Zhang, Y., Li, F., 2012. Effects of partial root-zone irrigation on physiology, fruit yield and quality and water use efficiency of tomato under different calcium levels. Agr Water Manage. 104, 89-94.

75. Zhou, S., Wei, C., Liao, C., Wu, H., 2008. Damage to DNA of effective microorganisms by heavy metals: Impact on wastewater treatment. J Environ Sci. 20, 1514-1518.
76. Zhou, Q., Li, K., Jun, X., Bo, L., 2009. Role and functions of beneficial microorganisms in sustainable aquaculture. Bioresource Technol. 100, 3780-3786.
77. Žanić, K., Dumičić, G., Škaljac, M., Ban, S.G., Urlić, B., 2011. The effects of nitrogen rate and the ratio of $NO3^-$:$NH4^+$ on Bemisia tabaci populations in hydroponic tomato crops. Crop Prot. 30, 228-233.

Chapter 5

# The Use of Glyphosate in Sugarcane: A Brazilian Experience

Carlos Alberto Mathias Azania[1],
Luciana Rossini Pinto[1], Rodrigo Cabral Adriano[1],
Dilermando Perecin[2], and Andréa Padua Azania[1]

[1]Instituto Agronômico de Campinas - IAC, Centro de Cana, Ribeirão Preto, Brazil

[2]Universidade Estadual Paulista – Unesp, Jaboticabal, Brazil

## INTRODUCTION

In Brazil, sugarcane (*Saccharum* spp.) fields are renewed in intervals of five to six profitable crops on average. With each harvest, sugarcane displays a decrease in productivity due to diverse factors. Genetic, phytosanitary and edaphoclimatic issues are the main factors contributing to the degeneration that necessitates the renewal of sugarcane fields with more productive cultivars. After the last economical harvest, the ratoon crop is destroyed using mechanical or chemical processes or a combination of both. Chemical destruction is more practical and causes less impact on soil structure and quality due to less soil disturbance. Glyphosate is the most widely used non-selective herbicide in the chemical eradication of ratoon crops because there is a broad spectrum of plants susceptible to glyphosate.

Glyphosate's mechanism of action is through inhibition of the 5-enolpyruvylshikimate-3-phosphate synthase (EPSPS), a precursor of the aromatic amino acids phenylalanine, tyrosine and tryptophan, which are essential for protein synthesis. The recommended dosage of glyphosate for the eradication of plants is 1440 to 2880 g acid equivalents (a. e.) $ha^{-1}$ [1,2]. However, sugarcane cultivars present varying degrees of susceptibility and require different amounts of herbicide for the complete death of the plant.

In Brazil, sugarcane cultivars commercially released by genetic improvement programmes are not characterised in terms of their susceptibility to glyphosate. Nevertheless, knowledge of the degree of cultivar tolerance to glyphosate can generate savings for producers and benefit to the environment through the reduction of the quantity of applied herbicide. Literature studies of cultivar responses to herbicides, especially glyphosate, are supported solely by phytotechnical observations, such as plant height, girth and mass gain. However, these characteristics are greatly affected by the environment and require longer periods of evaluation and data collection, such as the 12-month studies of [3] and [4].

The use of isoenzymatic markers allows for the prompt analysis of results with larger sample numbers while using a relatively simple and inexpensive technique that can substitute or reduce field experimentation. However, the choice of the correct enzymes to analyse is critical to the success of this technique and obtaining robust results.

# SUGARCANE CROPS

Sugarcane probably originated in New Guinea, and from there, it was introduced to India, where the oldest evidence of its existence has been recorded [5]. Officially, Martins Afonso de Souza brought the first sugarcane plant to Brazil in 1532 and started its cultivation in the Captaincy of São Vicente (Capitania de São Vicente). This transfer was the beginning of an industry that found in Brazil, among other nations that would later initiate production, its most fertile ground for rapid expansion and perpetuation for an almost uninterrupted 500 years. Starting in the 1970s, sugarcane farming became increasingly important for Brazil as the agro-industrial sector was tapped to contribute to a solution to the emerging energy crisis because of the potential for energy production from sugarcane as a renewable source [6]. Growth in the sugarcane-ethanol sector is important for the Brazilian economy in that the sector's growth entails both the creation of jobs and of 100% national renewable energy.

According to taxonomic classification, sugarcane belongs to the *Poaceae* family and the *Saccharum*genus. Sugarcane is a semi-perennial plant requiring a tropical or sub-tropical climate [7]. With a C4 metabolism, sugarcane is classified as having among the highest rates of photosynthetic efficiency and a high efficiency for water usage [8]. The sugarcane plant is divided into aerial (culm, leaves and inflorescences) and underground parts (roots and rhizomes). The culms are cylindrical and are composed of nodes and internodes; these parts are defined as the aboveground portion that supports the leaves and inflorescences [9]. According to [10], each node has one alternating bud and a root system. Inflorescences are panicles with a hermaphrodite flower containing one ovule; the pistils terminate in purple or reddish stigmae that characterise the flower's plumose panicle [9]. The root system is fasciculated and serves to support, as well as to absorb and transport water and nutrients [8]. Sugarcane tillering influences the sugarcane handling system because each tiller behaves as an independent plant with individual organs, such as roots, leaves and fruits [11].

The most appropriate agricultural conditions for sugarcane propagation are found between the 30° north and 30° south latitudes, which are characteristic of tropical and subtropical regions. Outside of these latitudes, lower temperatures limit the growth and development of the plant [12]. According to [10], the optimal temperature range for the growth of this crop is between 20 and 35 °C with an ideal photoperiod of 10 to 14 hours [12] and an annual rainfall ranging between 1,000 and 1,600 mm, preferentially with abundant rain during the vegetative growth period and a dry period during maturation, which favours increased sucrose accumulation [13].

In Brazilian regions where it is traditional to grow sugarcane, planting may occur at different times of the year, as long as the producer possesses an irrigation system and cultivars that are adapted to each season [14]. Traditionally, in south-central Brazil, there are two cycles for planting: "cane of the year" and "year and a half". In "cane of the year", planting is performed between September and November, and the cane is harvested after 12 months [14]. This type of plantation addresses the demand for raw materials in the spring cycle (at the end of the harvest). In "year and a half" cane, planting is performed between January and April-May. In contrast to cane of year, this cycle allows for harvest during the autumn season (the beginning of harvest). Additionally, several producing units have practiced winter planting, particularly June through July, using rescue irrigation, and these units have obtained great productivity compared to "year and a half" cane planting.

Currently, Brazil is the largest producer of sugarcane in the world followed by India, China and Thailand [15]. The national production is estimated as 641.982 million tons with an average productivity of approximately 76.4 t $ha^{-1}$ [16].The national sugar-energy industry sector accounted for 1,283,258 jobs up to 2008 with 37.5% occupied by plant growth, 44.8% in the production and refining of sugar and 17.7% in the production of ethanol. This sector also accounted for approximately 3.85 million people that are employed indirectly [17]. The production and processing of sugarcane is currently managed by the private sector in Brazil, which achieves the lowest cost for production worldwide for both sugar and ethanol, emerging as a highly competitive segment in international markets [18].

# CHARACTERISTICS OF THE HERBICIDE GLYPHOSATE

Glyphosate was commercially released in 1974 under the trade name Roundup initially in the USA for industrial purposes, in the United Kingdom for use in wheat crops and in Malaysia for use in rubber trees. Currently, the molecule is registered in more than 130 countries for the control of over 300 species of weeds in over 100 types of crops [19], making it the most widely used herbicide [20]. Worldwide, there are numerous registered trademarks of the herbicide, which, according to [50], number more than 150. In Brazil, glyphosate is also registered for the eradication of sugarcane ratoon crops [21].

The molecule belongs to the glycine-derived chemical group (Group G), and its mechanism of action consists of inhibiting the

5-enolpyruvylshikimate-3-phosphate synthase (EPSPS) in the shikimic acid pathway [22], which is found only in microorganisms and plants [23]. The molecular formula of glyphosate is $HO_2CCH_2NHCH_2PO(OH)_2$. Glyphosate has a solubility in water of 15,700 mg $L^{-1}$ at 25 °C and pH 7, a density of 1.74 g $mL^{-1}$, a vapour pressure of $2.45 \times 10^{-8}$ Pa (45 °C), $pK_a$ values of 2.6, 5.6 and 10.3 (acid) and a $k_{ow}$ between 0.0006 and 0.0017. In the soil, glyphosate is strongly adsorbed to colloids, and its leachability is notably low. The compound has an average $K_{oc}$ of 24,000 mL $g^{-1}$, and its volatilisation and photodegradation are negligible [22]. The half-life of the molecule in the environment depends on the surrounding soil texture and microbial activity and may vary from a few days to several years [24]. In roots, the absorption is slow due to the low diffusion and high adsorption to the soil, which also favours microbial action in the transformation of the molecule into its main metabolite, aminomethylphosphonic acid [25; 26]. Glyphosate is absorbed through leaf cuticles, and its translocation occurs mainly via the cellular symplast to the leaves and apical meristem, as well as to underground organs [22]. According to [27], glyphosate absorption depends on such factors as the age of the plant, environmental conditions, surfactants and herbicide concentration in the soil milieu.

As an herbicide, glyphosate is among the less hazardous agro-toxins used in agriculture [28]. Glyphosate-based herbicides, when used according to their respective guidelines, display low toxicity and are safe to humans [30].

Often, the glyphosate molecule is not efficient in penetrating waxy cuticles. Therefore, commercial formulations contain surfactants capable of reducing surface tension in herbicide droplets, thus increasing their penetration in leaves [20]. However, these surfactants are more toxic than the glyphosate molecule [29]. For example, polyoxyethylene amine, the predominant surfactant in Roundup® [30], has been classified as moderately to highly toxic in laboratory tests [31]. Glyphosate is a unique molecule, and although it is considered to be of low toxicity, its unrestrained use can affect the environment through direct or indirect effects on non-target organisms [32].

In plants, the EPSPS enzyme catalyses a reaction between shikimate-3-phosphate (S3P) and phosphoenolpyruvate (PEP) to produce 5-enolpyruvylshikimate-3-phosphate and inorganic phosphate. Glyphosate binds to the catalytic site of EPSPS and to the S3P substrate to form the EPSP synthase-S3P-glyphosate complex [33]. The relevance of the shikimate pathway is such that approximately 35% of all plant mass is related to derivatives from this pathway; moreover, 20% of all of the carbon fixed during photosynthesis also travels through this metabolic pathway [34].

According [35], the inhibition of amino acids compromises the production of carotenoids and chlorophyll, thus causing irreversible cellular damage. Therefore, the translocation of the herbicide throughout the entirety of the plant causes plant death in a few days or weeks (Figure 1). The inhibition of EPSPS leads to the accumulation of high levels of shikimate in vacuoles, which is intensified by the loss of control of the carbon flow across this pathway [36]. Thus, there is an obstruction in the production of the aromatic amino acids phenylalanine, tyrosine and tryptophan, which are essential for protein synthesis and serve as precursors for secondary metabolites that are important for plant growth [37], resulting in the slow development of symptoms [33].

# REFORMING SUGARCANE FIELDS AND THE USE OF GLYPHOSATE

With each harvest cut, sugarcane sprouts new tillers that develop into culms [38]. Nevertheless, ratoon-crop productivity gradually diminishes with an increasing number of cuts [39], thereby requiring the renewal of the field. In the state of São Paulo, the average productivity of sugarcane fields is approximately 80 to 85 t $ha^{-1}$, considering the longevity of ratoons to be between five and six cuts [40]. However, in the region of Ribeirão Preto, SP, ratoon crops after the sixth cut are no longer economically viable, and renewal of the sugarcane field is necessary [41]. Degeneration after successive years of production makes the renewal of sugarcane fields essential. The causes of degeneration are diverse and involve a combination of genetic, physiological, phytosanitary, edaphoclimatic and phytotechnical factors. The factors impacting degeneration may also be linked to characteristics of the growing environment, such as a decrease in soil fertility [42]. Another cause for degeneration can be soil compaction and consequent difficulties in root development, as proposed by [43].

The authors note that compacted soil still presents difficulties for root development, even if its humidity levels are close to the soil's capacity.

[44] found that degeneration is linked to the health of the plants. Sugarcane-field longevity may also be affected by competition with weeds [45], nematode infestations [46] and uprooting of tufts during mechanical harvesting [47].

The fact that various cuts are performed from a single plantation allows for the formation of a significant number of root systems, which often make the elimination of the ratoon crop difficult, especially if the eradication is performed mechanically, which may also compromise the settlement of the next plantation. At the time of crop renewal, the ratoon crop is first eliminated through desiccating herbicides, specifically glyphosate, and after plant death, eradication is later completed using mechanical destruction of the crop [48].

Glyphosate is the most widely used herbicide for the chemical eradication of sugarcane ratoons due to its ease of use, low cost and absence of residual effects on the soil, which allows for repeated plantation in the same area, as is often practiced by farmers [49, 50]. Tolerance to glyphosate is highly prevalent in cultivars, and while certain cultivars are eradicated with a dose of 1080 g a. e. $ha^{-1}$, others require a dose of 2520 g a. e. $ha^{-1}$. According to [1,2], the minimal lethal dose for sugarcane is 1440 g a. e. $ha^{-1}$.

The progression of symptoms caused by a glyphosate application occurs in a gradual fashion (Figure 2) until the eradication of the plants [23, 36]. The authors noted that glyphosate-induced damage develops slowly until complete death in contrast to the effects of other herbicides. According to these authors, molecular stability inside the plant allows for the occurrence of irreversible effects on processes that control both annual and perennial plants (Figures 3 and 4).

After the application of glyphosate in the eradication of sugarcane crops, there was a stunting of plant growth, with treated plants retaining the same size up to 45 days after herbicide application [51]. The negative effect on growth was evaluated by measuring the plant height, with the treated plants maintaining similar average height values throughout the evaluation period in contrast to controls, which were able to maintain vigorous vegetative growth. The growth stunting was due to the indirect influence of glyphosate on the regulators of plant growth, such as indole-3-acetic acid. This hormone is fundamental for cellular elongation, apical dominance and stem and root growth and is dependent on the shikimate pathway, being inhibited when there is a disruption of EPSPS [36].

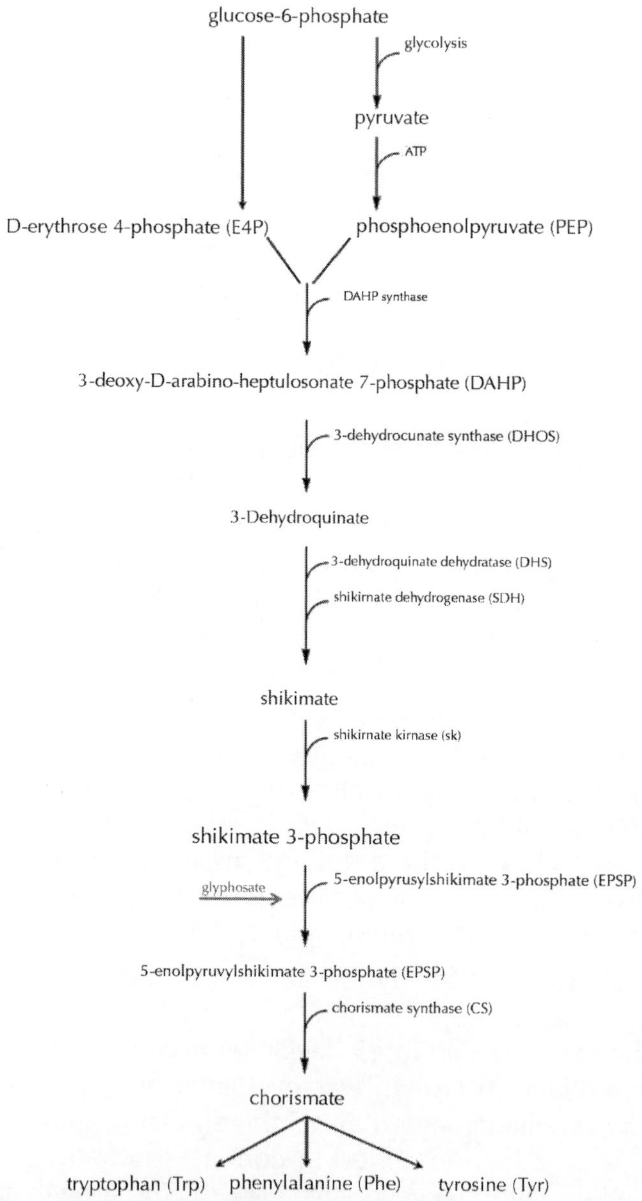


**Figure 1:** The shikimate pathway and the action of glyphosate on plants.

**Figure 2:** Sugarcane plants ten days after the application of glyphosate (2880 g a.e. $ha^{-1}$).

The varying tolerance of sugarcane cultivars to glyphosate was studied [52], who found different sensitivities among cultivars. The authors also classified the genotypes IAC86-2210, IAC83-1313, IAC82-2045, PO83-698 and IAC83-4157 as susceptible to glyphosate, IAC86-3154, IAC87-3184, RB72454 and SP80-1842 as of intermediate susceptibility, and IAC82-3092, IAC87-3396 and RB806043 as tolerant. Nevertheless, complete death, even in the less susceptible cultivars, occurred after 45 days following application. A plant's inherent tolerance is related to the plant's capability for absorption, translocation, metabolism and/or elimination of a herbicide [53]. In [54] also noted that differences in absorption depend primarily on morpho-anatomical characteristics of the species and that in the aerial parts of the plant, absorption is highly influenced by the presence or absence of cuticles. The physicochemical content of the leaf surface is another form of plant resistance to glyphosate [55]. According to these authors, leaves with flat cuticle surfaces and without large quantities of wax can better retain applied droplets. After penetration, the herbicide can then be metabolised into secondary compounds without herbicidal activity, or its potency might be enhanced [56].

The plant's development stage is another factor that should be considered in the eradication of cultivars because plants must be 40 to 80 cm tall at the time of glyphosate application [2], and the total leaf area must be sufficient to intercept the herbicide. The inherent resistance to glyphosate is greater in taller plants [57]. After the formation of culms, plants become more tolerant to the herbicide [58].

A relationship between plant size and glyphosate efficacy was also observed [59] while studying*Conyza bonariensis*. The authors observed that herbicidal efficacy was greater when the plants presented up to two pairs of leaves. Nevertheless, in more advanced stages of development, it was necessary to increase herbicide dosage by up to fivefold.

Glyphosate applied 40 days after the last harvest caused the highest percentage of dead tillers, and also, genotypes IAC87-3184, RB835489 and SP87-344 displayed high to intermediate sensitivity, while IAC91-5155 was considered tolerant to the herbicide [49]. The most effective application time for eradication was 65 days after cane harvesting [60]. The same authors reported that a dose of 960 g a. e. ha$^{-1}$ eradicated the majority of cultivars, except for Co997, which needed 1920 g a. e. ha$^{-1}$ of the herbicide.

**Figure 3:** Intoxication symptoms caused by glyphosate rates in sugarcane cultivars (IACSP94-2094-4004 and IACSP94-4004) to 27 days after application. Instituto Agronômico de Campinas -IAC, 2012.

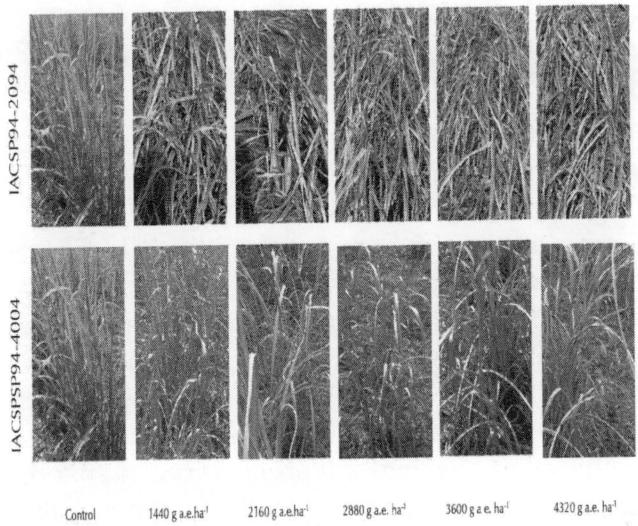


**Figure 4:** Intoxication symptoms caused by glyphosate rates in sugarcane cultivars (IACSP94-2094-4004 and IACSP94-2191) to 45 days after application. Instituto Agronômico de Campinas - IAC, 2012.

The development of transgenic sugarcane plants, particularly those with tolerance to glyphosate, will most certainly change the way in which sugarcane is eradicated. The use of glyphosate will reduce the costs associated with the control of weeds across cycles; however, during the period of cane eradication, the herbicide will now have a limited impact due to the tolerance introduced to the cultivars. In this case, eradication may have to be performed mechanically, which will have a negative impact on soil conservation and might stimulate weed germination from soil propagule banks. From this perspective, it is important to emphasise that research aimed at sugarcane plants tolerant to glyphosate should also consider the use of herbicides in the eradication of future cultivars.

In the eradication of ratoon crops, glyphosate is used to eradicate the crop and also to control emerged weeds. However, when a sugarcane field also possess weeds that are hard to control, such as*Cynodon dactylon* and *Cyperus rotundus*, the use of higher doses of residual herbicides after the application of glyphosate is adopted in a process known as "disinfestation".

The aggressive biological nature of hard to control weed species requires that handling start with the desiccation of the plants to optimise the use of glyphosate in the eradication of ratoon while also being able to introduce residual herbicides at higher doses. In these instances, the use of glyphosate in crop eradication serves the dual role desiccating the ratoons and controlling problematic weeds.

In the time period following eradication but before planting sugarcane, the producer should formulate a strategy and opt for techniques that ensure higher sustainability of the system. These methods include such techniques as crop rotation or planting green-manure crops, although these may still compromise the techniques' sustainability if installed in fields that have been infested with "difficult-to-control weeds" or previously treated with residual herbicides for "disinfestation". The producer should carefully plan to use techniques that generate the most effective soil preparation and handling of weeds while simultaneously ensuring that after the treatments, the soil remains prepared for a new sugarcane plantation.

# THE SHIKIMATE PATHWAY AND ISOENZYMATIC MARKERS

The shikimate pathway is found only in plants and microorganisms and is completely absent in mammals, birds, reptiles, fish and insects. These organisms extract the aromatic compounds necessary for survival and reproduction from their diet, while plants must produce such compounds because they do not have alternative means to obtain the compounds [23].The shikimate pathway is initiated with the reaction of PEP and erythrose 4-phosphate, a reaction catalysed by the enzyme DAHP synthase (3-deoxy-D-arabino-heptulosonate-7-phosphate synthase [61]. The resulting product is the seven-carbon acyclic intermediate 3-deoxy-D-arabino-heptulosonate-7-phosphate (DAHP).

DAHP is converted to cyclic form through catalysis by 3-dehydroquinate synthase in the presence of $NAD^+$ as a coenzyme. In this process, 3-dehydroshikimate dehydratase dehydrates the cyclic form of DAHP. Next, shikimate dehydrogenase, in the presence of $NADP^+$ (oxidised NADPH), reduces the cyclic and dehydrated DAHP to shikimate. The molecule is later phosphorylated by the SP3 kinase, which converts a molecule of ATP to ADP. The phosphorylated shikimate subsequently reacts with one molecule of PEP in the presence of 5-enolpyruvylshikimate-3-phosphate (EPSP) synthase, resulting in the production of EPSP. The shikimate pathway terminates with the production of chorismate (chorismic acid) through the dephosphorylation of EPSP by chorismate synthase.

Other metabolites essential to plant life may be produced from chorismate, including the amino acids tryptophan, phenylalanine and tyrosine, as well as vitamin K, ubiquinone and tetrahydrofolate [62, 23]. The amino acid phenylalanine is a precursor not only of proteins but also of other secondary products, such as phenolic compounds, anthocyanins, lignin and promoters and growth inhibitors. Tryptophan is also a precursor to indole-3-acetic acid, which is responsible for apical dominance and is vital for cellular growth and several other regulatory processes. Therefore, inhibitors of the shikimate metabolic pathway represent a strategic alternative in the development of herbicides with low environmental impact, such as glyphosate [23]. In this context, it could be interesting to use protein electrophoresis as a tool to study the eradication of plants by glyphosate using the isoenzymes involved in the metabolic pathway of shikimic acid.

Isoenzymes are the multiple molecular forms of enzymes that perform the same or similar catalytic activities. These enzymes are coded by one or more genes and may play an important role in survival across diverse environments. Isoenzymes are directly affected by both biotic and abiotic stressors [63]. The band intensity and isoenzymatic profile are plant-, tissue- and development stage-specific [64]. Some factors that affect plant metabolism, such as mineral nutrition, low temperature and diseases, among others, influence the activity of isoenzymes, specifically, esterases, peroxidases, phosphatases and phenolases, which in turn generate different expression patterns and levels of activity.Isoenzymatic patterns were used as tools by [65] who concluded that the enzymatic system of malate dehydrogenase is an efficient marker for aerobic respiration in pepper seeds during the maturation period. In Serbia [66] (the Vojvodina region), observed that the shikimate dehydrogenase system is also an efficient isoenzymatic marker for the study of genetic variability and polymorphisms in different almond genotypes.

In [67] was used the same technique in soy cultivars and registered a difference in electrophoretograms in terms of peroxidase activity. In

[68] used isoenzymatic markers to identify species of lettuce nematodes, and [69] used the method "in vitro" in sugarcane to observe varietal differences among doses of glyphosate.

# ESTERASE ISOENZYMES IN ABIOTIC STRESS

Enzymatic activity is influenced by stress factors, such as non-optimal temperature or nutrient levels and infection by pathogens. These stress responses subsequently lead to gene activation and, as a consequence, to the emergence of several molecular forms [70]. Because of the involvement of isoenzymes in changes to metabolism and defence mechanisms in plants, studies involving isoenzymes can be used in cases of both biotic and abiotic stress [63]. The authors report that polymorphisms displayed by isoenzymes are intermediate products of gene expression and are closer to the final phenotypical expression than those of DNA polymorphisms.

Esterases are isoenzymes comprising a group of genetically distinct enzymes that are found across a large spectrum of living organisms and that play a large variety of roles; nevertheless, esterases display a common trait of catalysing the hydrolysis of esters, peptides, amides and halide bonds [71]. Esterases can be found as both monomers and dimers [72, 73]. Esterases are significantly linked to lipid metabolism, such as that of membrane phospholipids, due to catalysis of ester hydrolysis [74].

In polyacrylamide gel electrophoresis assays, the α-esterase isoenzyme is detected using naphthyl ester substrates and histochemical stains. Enzyme isoforms with an affinity to hydrolyse α-naphthyl acetate are identified on the gel as black bands derived from the precipitation of α-naphthol, which results from the hydrolysis of α-naphthyl with the Fast blue RR salt stain [75].

Esterase isoenzymes have also been extensively explored in studies of genetic diversity due to their high rate of polymorphisms [76, 77]. In [78] the authors reported that sugarcane cultivars could be identified using esterase isoenzymes, and [79] in studying the parameters for sugarcane differentiation, observed that the electrophoretic profile of esterases is maintained in plants of varying physiological ages, as long as the growth environment is controlled.

Esterase isoenzymes are among the most widely used enzymes in the evaluation of enzymatic alteration in plants that are affected by parasitic nematodes across various pathosystems [80]. In [75] was studied esterase polymorphisms in 16 cultivars of soy that underwent or were spared treatment with glyphosate. The authors observed variation in the sensitivity to α-esterase isoenzymes of the different cultivars and also found that sensitivity did not seem to be connected with the homozygous RR status of the genetically modified plants.

## PRACTICAL RESULTS OF THE ISOENZYMATIC PROFILES OF SHIKIMATE DEHYDROGENASE AND A-ESTERASE IN SUGARCANE

The tolerance of sugarcane cultivars to chemical eradication using varying doses of glyphosate was investigated [51] using phytotechnical parameters and isoenzymatic markers. The author hypothesised that the study of isoenzymatic profiles of shikimate dehydrogenase and α-esterase could optimise phytotechnical fieldwork observations regarding herbicidal tolerance. Shikimate dehydrogenase was selected because it is involved in the shikimic acid pathway, which is affected by glyphosate action, and α-esterase was chosen because it is associated with oxidative stress. The isoenzymatic profiles of shikimate dehydrogenase and α-esterase were studied in sugarcane cultivars IACSP94-2094, IACSP94-2101, IACSP93-3046, IACSP94-4004, IAC86-2480 and RB72454 at 8, 24, 48, 72 and 144 hours after the application of glyphosate at doses of 0, 1440, 2160, 2880, 3600 and 4320 g a. e. $ha^{-1}$. The results showed that the bands for shikimate dehydrogenase tended to position near the cathode (at the top of the gel), while α-esterases were positioned closer to the anode, due to a greater migration during the gel run. The enzymatic system of shikimate dehydrogenase presented bands that were less sharp and also had a lower number of bands (three). There were no observed polymorphisms among cultivars, regardless of whether the data were analysed according to herbicide dosages or in relation to controls. Therefore, the isoforms remained constant among the different cultivars and treatments. In [81] was studied 20 enzymatic systems in the identification of sugarcane cultivars and also did not obtain any promising results using shikimate dehydrogenase. In [82] was studied populations of *Stryphnodendron adstringens*, known in Brazil as *barbatimão,* and also did not find polymorphisms for shikimate dehydrogenase.

The enzymatic system of α-esterase was specific for each studied cultivar, allowing for cultivar identification based on this biochemical marker [51]. This observation corroborates the findings [83], who created an analytical key for sugarcane cultivars and found a different pattern of α-esterase in each of the ten cultivars studied. A large number of bands of the α-esterase enzymatic complex were found with varied band intensity and thickness. The characteristics of this complex can be related to the degree of ploidy of the plant species; sugarcane is polyploid [70, 84].

Cultivars of variety IACSP93-3046 and RB72454 did not present differences in terms of bands owing to the application of glyphosate. These cultivars were considered susceptible to glyphosate based on field studies reporting a percentage of tiller death of 93.16% and 94.25% respectively. Moreover, marked toxic effects were rated as high as 94% for IACSP93-3046 and 95.5 % for RB72454 [51].

Sugarcane cultivar IACSP94-4004 was the only cultivar to show an alteration in its band pattern due to the application of glyphosate. Across all of the treatments in which the herbicide was applied, there were two additional bands that were not present in the controls and that were present from the first assessment 8 hours after the application (HAA) of herbicide to the last assessment at 144 HAA. The appearance of additional bands may be due to the expression of genes from this enzymatic system in response to stress caused by glyphosate treatment, thereby demonstrating that the cultivar response to the herbicide is directly linked to the genotype of each cultivar variety. In fact, cultivar IACSP94-4004 was relatively tolerant to the field experiments, as it was the cultivar to show the lowest average (percentage) of tiller death at 45 HAA (80.15 %), the symptoms of toxicity in this cultivar were less pronounced with an average of 82.5 %, and only glyphosate doses of 3600 g a. e. $ha^{-1}$ caused symptoms similar to those caused by the highest dose of 4320 g a. e. $ha^{-1}$.

Evaluation of the isoenzymatic system of esterases has been used in other studies to characterise cultivar tolerance. Nevertheless, in the evaluation of sugarcane eradication, the isoenzymatic profiles of shikimate dehydrogenase and α-esterase did not constitute a reliable tool [51].

## THE USE OF GLYPHOSATE IN SUGARCANE AS A RIPENER

Maturation is one of the most important aspects of sugarcane crops because maturation is directly related to the optimal time-point for harvest/industrial transformation.For the plant to enter the maturation process naturally, one or more sources of stress are necessary, with a gradual reduction in photoperiod, temperature or precipitation being the most effective stimulants [85]. In Brazil, more specifically in the south-central region, the maturation process is initiated in April/May when the climate becomes colder and drier. However, even under favourable conditions, sugarcane maturation may also be induced in responsive cultivars as a strategy to produce high-quality raw material across all of the different phases of harvesting.

To guarantee that maturation is complete, uniform, early and programmed and to avoid undesirable flowering, the sugarcane-ethanol industry has been adopting the use of ripeners (growth regulators) in sugarcane. Ripeners are chemical compounds that induce the translocation and storage of sugars, mainly sucrose, in the culm. Therefore, the goal of the ripeners is both to advance and maintain natural maturation and to provide high-quality raw materials for early industrial transformation, as well as to aid in the handling of cultivars [85].

The same authors state that to artificially induce maturation, growth regulators are applied by aircraft eight to ten months after the last harvest, that is, during the plants' vegetative state. In practice, the months of February and March or October are the periods during which farmers aim to apply the enhancers because they can anticipate the beginning or the end of the harvest.

There are two basic types of ripeners for sugarcane fields, "non-stressors" and "stressors". Non-stressor ripeners do not diminish the plants' growth rate, and their action induces the release of ethylene, the compound responsible for maturation that helps in the accumulation of sucrose in sugarcane culms. Stressor compounds, such as glyphosate, are growth inhibitors that markedly decrease the sugarcane growth rate, making the plants accumulate sucrose instead of expending it as an energy source for growth. This reduction in growth rate forces the plant to mature [85]. In [86] sugarcane plants with stagnated growth stop sprouting new leaves, and as a consequence, the reduced number of phytochromes in chloroplasts becomes insufficient to detect the photoperiod and thus stimulate the transition of the apical bud from vegetative to reproductive.

The effect of glyphosate, after it is applied to sugarcane, has a rapid onset, allowing for an increase of sucrose accumulation in 30 to 40 days after application. The glyphosate dose used is normally 0.3 to 0.4 l $ha^{-1}$ and may reach a maximum of 0.6 – 0.8 l $ha^{-1}$, leading to differences in maturation speed as a function of dosage. Harvest should be performed when the highest levels of sucrose are reached [12].

# CONCLUSIONS

Glyphosate is a high-efficiency molecule in the sugar-ethanol industrial sector in Brazil. Glyphosate can be used as an herbicide in the control of weeds and the chemical eradication of sugarcane crops, and when applied in low dosages, glyphosate can be used as a ripener.

Independently of where the product is used in the production system, the key factor is the adequacy of the dosage. If cultivar tolerance is known, it is possible to adjust the dose of glyphosate to eradicate sugarcane ratoon crops. Research has shown that a dose of 2 kg $ha^{-1}$ (1440 g a. e. $ha^{-1}$) of the commercial product Roundup WG eradicates 92.76% of the tillers of the IACSP94-2094 cultivar but only 40.3% of the IACSP94-4004 cultivar. For the most effective eradication of the IACSP94-4004 cultivar, 4 kg $ha^{1}$ (2880 g a. e. $ha^{-1}$) of chemical is needed. This finding demonstrates that knowledge of plant tolerance can be a valuable tool to adjust glyphosate doses to the appropriate concentrations for ratoon-crop eradication.

Reducing the quantity of applied glyphosate is possible, as long as the crop is sensitive. Environmental and economic benefits can also be obtained by applying lower quantities of the herbicide. Information on cultivar tolerance also allows producers to know when to use higher herbicide concentrations than those recommended in the literature. However, if the producer applies a lower dose than is necessary, the ratoon crop will not be completely eradicated. As a consequence, there will be additional expenditure on a additional herbicide application or on mechanical eradication, which is highly problematic because it causes the greatest disruption to the soil and later leads to the presence of crop stubble in the new sugarcane field.

# ACKNOWLEDGEMENTS

The authors are grateful to FAPESP (Fundação de Amparo à Pesquisa do Estado de São Paulo) for the financial support (Grant # 2010/09016-9).

# REFERENCES

1. Bacchi OOS, Rolim JC. Dose letal de glifosato para fins de eliminação química de soqueiras. Proceedings of the 2nd Congresso Nacional da Sociedade dos Técnicos Açucareiros e Alcooleiros do Brasil; 1981, Rio de Janeiro. Rio de Janeiro: STAB, 1981. p.13-20.
2. Lorenzi H. Pragas da cultura da cana-de-açúcar. Proceedings of 2nd Reunião Técnica Agronômica, 1983, Piracicaba. São Paulo: Copersucar, p. 59-82.
3. Souza Jr, Perecin D, Azania CAM, Schiavetto AR, Pizzo IV, Candido LS. Tolerância de cultivares de cana-de-açúcar a herbicidas aplicados em pós-emergência. Bragantia 2009; (68): 941-951.
4. Zera FS, Azania CAM, Schiavetto AR, Lorenzato CM, Freitas GB, Azania AAPM. Tolerância de mamona (*Ricinus communis*) a herbicidas utilizados na cultura da cana-de-açúcar. Nucleus 2011 (8): 453-462.
5. Machado FBP. Brasil, a doce terra. http://www.canaweb.com.br/conteudo/Historiadosetor.htm/ (accessed 9 February 2001).
6. Kuva MA. Efeitos de períodos de controle e de convivência das plantas daninhas na cultura da cana-de-açúcar (*Saccharum* sp) no Estado de São Paulo [dissertation]. Universidade de São Paulo, Piracicaba; 1999.

7. Vidal RA, Trezzi MM. Origem da cultura e sua importância. In: Teoria e prática do manejo de infestantes na cultura da cana-de-açúcar no Brasil. Porto Alegre: 2011.
8. Casagrande AA, Vasconcelos ACM. Fisiologia da parte aérea. In: Dinardo-Miranda LL, Vasconcelos ACM, Landell MGA (ed.). Cana-de-Açúcar. Campinas: Instituto Agronômico; 2008. p. 57-78.
9. Scarpari MS, Beauclair EGF. Anatomia e botânica. In: Dinardo-Miranda LL, Vasconcelos, ACM, Landell MGA (ed.). Cana-de-Açúcar. Campinas: Instituto Agronômico; 2008, p. 45-56.
10. Casagrande AA. Tópicos de morfologia e fisiologia da cana-de-açúcar. Jaboticabal: FUNEP; 1991.
11. Alexander AG. Sugarcane physiology: a comprehensive study of the Saccharum source-to-sink system. Amsterdam: Elsevier; 1973.
12. Rodrigues JD. Fisiologia da cana-de-açúcar. Botucatu: Fepaf; 1995.
13. Agrianual 2007: Anuário da agricultura brasileira. São Paulo: Instituto FNP, 2007. p.23-28.
14. Anjos IA, Figueiredo PAM. Aspectos fitotécnicos do plantio. In: Dinardo-Miranda LL, Vasconcelos ACM, Landell MGA (ed). Cana-de-Açúcar. Campinas: Instituto Agronômico; 2008. p. 585 - 598.
15. FAO 2009 - Food and Agriculture Organization of the United Nations. http://faostat.fao.org/site/339/default.aspx/ (accessed 28 August 2011).
16. CONAB - Companhia Nacional de Abastecimento. Acompanhamento de safra brasileira: cana-de-açúcar, primeiro levantamento, maio 2011 – Brasília: Conab, 2011. http://www.conab.gov.br/ (accessed 28 de August 2011).
17. UNICA - Portal da União da Agroindústria Canavieira. Relatório de Sustentabilidade 2010. São Paulo, 2010. http://www.unica.com.br/ (accessed 10 August 2010).
18. Gonçalves DB. Dilemas do desenvolvimento sustentável na produção canavieira paulista. PhD thesis. Universidade Federal de São Carlos; 2005.

19. Halter S. História do herbicida agrícola glyphosate. In: Velini ED, Meschede D, Carbonari CA, Trindade MLB. Glyphosate (ed). Botucatu: Fepaf; 2009. p.11-16.
20. Jones DK, Hammond JI, Relyea RA. Competitive stress can make the herbicide Roundup® more deadly to larval amphibians. Environl Toxicol Chem 2011; (30):446–454.
21. AGROFIT – Sistema de agrotóxicos fitossanitários. Base de Dados. http://extranet.agricultura.gov.br/agrofit_cons/principal_agrofit_cons. 2011/ (accessed 20 February 2012).
22. Rodrigues BN, Almeida FS. Guia de herbicidas. 6th ed. Londrina: IAPAR; 2011.
23. Gruys KJ, Sikorski JA. Inhibitors of tryptophan, phenylalanineand tyrosine biosynthesis as herbicides. In: Singh BK (ed.). Plant amino acids: biochemistry and biotechnology. New York: Marcel Dekker, 1999. p. 357-384.
24. Arantes SACM, Lavorenti A, Tornisielo, VL. Efeito da calagem na mineralização de 14C-glifosato em solos. Ciência e Agrotecnologia 2011; 35 (2) 234-241.
25. Franz JE, Mao, MK, Sikorski JA. Glyphosate's molecular mode of action. In: Glyphosate. A unique global herbicide. American Chemical Society. Monograph 189, p. 521-615, 1997.
26. Galli AJB, Montezuma MC. Alguns aspectos da utilização do herbicida glifosato na agricultura. ACADCOM Editora; 2005. 67p.
27. Monquero PA, Christoffoleti PJ, Osuna MD, De Prado RA. Absorção, translocação e metabolismo do glyphosate por plantas tolerantes e suscetíveis a este herbicida. Planta Daninha 2004; (22): 445-451.
28. Duke SO, Powles SB. Glyphosate-resistant crops and weeds: now and in the future. AgBioForum 2009; (12): 346–357.
29. Dill GM, Sammons RD, Feng PCC, Kohn F, Kretzmer K, Mehrsheikh A, et al. Glyphosate: discovery, development, applications, and properties. In: Nandula, V. K. (ed.) Glyphosate resistance in crops and weeds: history, development, and management. Hoboken: Wiley; 2010. 321p.
30. Williams GM, Kroes R, Munro IC. Safety evaluation and risk assessment of the herbicide roundup and its active ingredient,

glyphosate, for humans. Regul Toxicol Pharmacol 2000; 31 (2) 117-165.

31. Bernal MH, Solomon KR, Carrasquilla G. Toxicity of formulated glyphosate (Glyphos) and Cosmo-Flux to larval Colombian frogs 1. Laboratory acute toxicity. J Toxicol Environ Health A 2009; (72) 961–965.
32. Reis MR. Impacto do glyphosate associado ao endossulfan e tebuconazole sobre microbiota do solo na cultura da soja. PhD thesis. Universidade Federal de Viçosa; 2009.
33. Duke SO, Powles SB. Glyphosate: a once-in-a-century herbicide. Pest Manag Sci 2008; (64) 319–325.
34. Kruse ND, Trezzi M, VIDAL RA. Herbicidas Inibidores da EPSPs: Revisão de literatura. Revista Brasileira de Herbicidas 2000; 1(2): 139-46.
35. Silva MD, Peralba MCR, Mattos MLT. Determinação de glifosato e ácido aminometilfosfônico em águas superficiais do Arroio Passo do Pilão. Revista Ecotoxicologia e Meio Ambiente, 2003 (13): 19-28.
36. Velini ED. Modo de ação do glyphosate. In: Velini ED, Meschede D, Carbonari CA, Trindade MLB (ed.). Glyphosate. Botucatu: Fepaf, 2009, 439p.
37. Tzin V, Galili G. New insights into the shikimate and aromatic amino acids biosynthesis pathways in plants. Mol Plant 2010; 3(6): 956–972.
38. Silva MA, Jeronimo EM, Dal'col LA. Perfilhamento e produtividade de cana-de-açúcar com diferentes alturas de corte e épocas de colheita. Pesqui Agropecu Bras 2008; 43(8): 979-986.
39. Sugawara LM, Rudorff BFT, Freitas CC, Picoli MCA, Adami M. Estimativa de produtividade de cana-de-açúcar (Saccharum officinarum L.) por meio de técnicas de análise de regressão linear múltipla. In: Proceedings of 13rd Simpósio Brasileiro de Sensoriamento Remoto; 2007; Florianópolis: INPE, p.435-442.
40. Yoshinaga EMS. As políticas de exploração da cana-de-açúcar no Brasil: da ocupação colonial a produção sucroalcooleira moderna [dissertation]. Universidade São Marcos, São Paulo; 2006.

41. Borba MMZ, Bazzo AM. Estudo econômico do ciclo produtivo da cana-de-açúcar para reforma de canavial, em área de fornecedor do estado de São Paulo. In: 47th Congresso Sober - Sociedade Brasileira de Economia Administração e Sociologia Rural, 2009, Porto Alegre.
42. King NJ, Mungomery RW, Hugues CG. Manual of cane growing. New York: Elsevier, 1965. 375 p.
43. Pacheco EP, Cantalice JRB. Compressibilidade, resistência a penetração e intervalo hídrico ótimo de um Argissolo Amarelo cultivado com cana-de-açúcar nos Tabuleiros Costeiros de Alagoas. Rev. Bras. Ciên Solo 2011; 35 (2): 403-415.
44. Bassinelo AI, Abrahão IS, Valadão MB, Barcellos JET, Piccolo CR. Primeiros resultados de estudos de novas variedades de cana-de-açúcar em solos de cerrado. In: Proceedings of 3rd Congresso Nacional da Stab and 5th Convenção da Actalac; 1984, São Paulo; 1984. p. 206-214.
45. Kuva MA, Gravena R, Pitelli RA, Christoffoleti PJ, Alves PLCA. Períodos de interferência das plantas daninhas na cultura da cana-de-açúcar. III - Capim-braquiária (Brachiaria decumbens) e capim-colonião (Brachiaria decumbens). Planta Daninha 2003; 21 (1): 37-44.
46. Dinardo-Miranda LL. Manejo de nematoides e pragas de solo em cana-de-açúcar. In: Campos AP, Vale DW, Araújo ES, Corradi MM, Yamauti MS, Fernandes OA, Freitas S (ed.). Manejo Integrado de Pragas. Jaboticabal: FUNEP, 2006. p. 59-80.
47. Salvi JV, Matos MA, Milan M. Avaliação do desempenho de dispositivo de corte de base de colhedora de cana-de-açúcar. Engenharia Agrícola 2007; 27 (1): 201-209.
48. Coleti JT. O preparo de solo sob a ótica conservacionista. In: Dinardo-Miranda LL, Vasconcelos ACM, Landell MGA (ed.). Cana-de-açúcar. Campinas: Instituto Agronômico, 2008. p. 573-584.
49. Silva MA, Carlin SD, Caputo MM. Tipos de colheita e épocas de aplicação de glifosato na erradicação de soqueiras de cana-de-açúcar. Pesqui Agropecu Bras, 2006; 41 (1): 43-49.

50. Galli AJB. A molécula glyphosate e a agricultura brasileira. In: Velini ED, Meschede DK, Carbonari CA, Trindade MLB (ed.). Glyphosate. Botucatu: Fepaf, 2009.p. 439.
51. Adriano RC. Características fitotécnicas e isoenzimáticas em cana-de-açúcar após a aplicação de glyphosate. [dissertation]. Instituto Agronômico, Campinas; 2012.
52. Silva MA, Rossetto R. Diferenças varietais na eliminação química de soqueiras de cana-de-açúcar. STAB, Açúcar, Álcool e Subprodutos 2002; (20): 24-27.
53. Christoffoleti PJ. Aspectos da resistência de Plantas Daninhas a Herbicidas. 3rd ed. Campinas: Associação Brasileira de Ação a Resistência de Plantas Daninhas aos Herbicidas (HRAC-BR), 2008. 120p.
54. Ferreira EA. Manejo de plantas daninhas tolerantes ou resistentes ao glyphosate no Brasil. In: Vargas L. (Ed.). Glyphosate: passado, presente e futuro. São Paulo: SBCPD, 2008.
55. Heredia A, Casado CG, Laguna L, Reina LL, Serrano JM, Domínguez E. La cutícula vegetal: estrutura y funciones. Ecología 1998; (12) 293-305.
56. Roman ES. Como funcionam os herbicidas: da biologia a aplicação. Passo Fundo: Gráfica Editora Berthier; 2007. 160p.
57. Ferreira MO. A eficiência do glifosato na destruição química das socarias de cana-de-açúcar das variedades NA56-79 e CB45-3. STAB, Açúcar, Álcool e Subprodutos 1986; (4): 46-48.
58. Procópio SO, Silva AA, Vargas L, Ferreira FA. Manejo de plantas daninhas na cultura da cana-de-açúcar. Viçosa: Editora Viçosa, 2003. 150p.
59. Dinelli G, Marotti I, Bonetti A, Minelli M, Catizone P, Barnes J. Physiological and molecular insights on the mechanisms of resistence in Conyza Canadensis (L.) Cronq biotypes. Pestic Biochem Physiol 2006; (86): 30-41.
60. Santos AJR, Graciano SP, Bacchi OOS, Kashiwakura Y. Doses e épocas de aplicação de glifosato na erradicação de soqueiras de diferentes variedades de cana-de-açúcar. In: Proceedings of the 3rd Congresso Nacional da Sociedade dos Técnicos Açucareiros e Alcooleiros do Brasil, 1984, São Paulo: STAB; p.276-281.

61. Lehninger AL, Nelson DL, COX MM. 2nd ed. São Paulo; Sarvier; 1995.
62. Devine M, Duke SO, Fedtke C. Inhibition of amino acid biosynthesis. In: Physiology of herbicide action. 1993. p. 251-294.
63. Torggler MGF, Contel EPB, Torggler SP. Isoenzimas – variabilidade genética em plantas. Ribeirão Preto. Sociedade Brasileira de Genética. 1995.
64. Peirce LC, Brewbaker JL. Applications of isozyme analysis in horticultural science. HortScience 1973; (8): 17-22.
65. Vidigal DS de, Dias DCFS, Von-Pinho EVR de, Dias LAS dos. Alterações fisiológicas e enzimáticas durante a maturação de sementes de pimenta (Capsicum annum L.). Rev Bras Sementes 2009; 31(2): 129-136.
66. Colich S, Milatovich D, Nikolich D, Zec G. Dehydrogenase isoenzyme polymorphism in selected almond genotypes (Prusus Amygdalus Batsch.). Bulg J Agric Sci 2009; (15): 552–556.
67. Menezes SM de, Tillmann MAA, Dode LB, Villela FA. Detecção de soja geneticamente modificada tolerante ao glifosato por métodos baseados na atividade de enzimas. Rev Bras Sementes 2004; 26 (2): 150-155.
68. Rabello LKC. Quantificação de danos e perdas causados por Meloidogyne spp. em alface (*Lactuca sativa* L.) [dissertation]. Universidade Federal do Espírito Santo; 2010.
69. Zambrano AY, Demey JR, González V. Selección in vitro de líneas celulares de caña de azúcar resistentes a glifosato®. Agronomía Tropical 2002; 52 (2): 139-160.
70. Gottlieb LD. Conservation and duplications of isozymes in plants. Science 1982; (216): 73-380.
71. Walker CH, Mackness MI. Estererases: problems of identification and classification. Biochemical Pharmacology 1983; (32): 3265-3269.
72. Brune W, Alfenas AC, Junghans TG. Identificações específicas de enzimas em géis. In: Alfenas AC. (ed). Eletroforese e marcadores bioquímicos em plantas e microrganismos. 2nd ed. Viçosa: Editora UFV; 2006. p. 202-328.

73. Weeden NF, Wendel JF. Genetics of plant isozymes. In: Soltis, D.E. & Soltis, P.S. (eds.). Isozymes in plant biology. London: Champman and Hall. 1990. p. 46-72.

74. Santos CMR, Menezes NL, Vilella FA. Modificações fisiológicas e bioquímicas em sementes de feijão no armazenamento. Rev Bras Sementes 2005; (27): 104-114.

75. Valentini LC. Caracterização genética e funcional de α- e β-esterases em cultivares de soja (Glycine max l. merrill) no estado do Paraná. [dissertation]. Universidade Estadual de Maringá; 2007.

76. Lopes RC, Casali VWD, Barbosa LCA, Cecon PR. Caracterização isoenzimática de oito acessos de erva-de-bicho. Horticultura Brasileira 2003; 21 (3): 433-437.

77. Martins CC, Bovi MLA, Mori ES, Nakagawa J. Isoenzimas na diferenciação de sementes de três espécies do gênero Euterpe. Revista Árvore 2007; 31 (1): 51-57.

78. Thom M, Maretzki A. Peroxidase and esterase isozymes in Hawaiian sugarcane. Hawaiian Planter›s Record 1970; 58 (6): 81-94.

79. Gonçalves CHRP, Cury JA, Crocomo OJ. Isoenzimas na identificação e seleção de variedades de cana-de-açúcar (Saccharum spp.). In: Proceedings of 2nd Congresso da Sociedade de Técnicos Açucareiros do Brasil; 1981. p. 329-340.

80. Medeiros JE. Seleção de bactérias para controle da meloidoginose e atividade isoenzimática de meloeiro parasitado por Meloidogyne incógnita. [dissertation]. Universidade Federal Rural de Pernambuco; 2007.

81. Gallacher DJ, Lee DJ, Berding N. Use of isozyme phenotypes for rapid discrimination among sugarcane clones. Aust J Agric Res 1995; (46): 601–609.

82. Glasenapp JS. Estrutura genética e fenóis totais de populações naturais de barbatimão (Stryphnodendron adstringens) [dissertation] Universidade Federal de Viçosa; 2007.

83. Almeida M, Crócomo OJ. Caracterização bioquímica de cultivares de cana-de-açúcar (*Saccharum* spp.): isoenzimas, proteínas solúveis e valor brix. Sci Agríc 1994; 51 (3): 422-429.

84. Kajihara D. Caracterização dos genes mustang em gramíneas com ênfase no estudo funcional em cana-de-açúcar [dissertation]. Universidade de São Paulo; 2008.
85. Bueno ACR, Machado DFSP, Azania AAPM, Azania CAM. Maturação estimulada. Revista Cultivar 2011; 12 (139): 30-32.
86. Araldi R, Silva FML, Ono EO, Rodrigues JD. Florescimento em cana-de-açúcar. Ciência Rural 2010; 40 (3): 694-702.

Chapter 6

# Metal Contamination of Soils and Prospects of Phytoremediation in and Around River Yamuna: a Case Study from North-central India

Manoj S. Paul[1], Mayank Varun[1], Rohan D'Souza[1], Paulo J.C. Favas[2, 4], and João Pratas[3], [4]

[1]Department of Botany, St. John's College, Agra, India

[2]Department of Geology, School of Life Sciences and the Environment, University of Trás-os-Montes e Alto Douro, Vila Real, Portugal

[3]Department of Earth Sciences, Faculty of Sciences and Technology, University of Coimbra, Coimbra, Portugal

[4]IMAR-CMA Marine and Environmental Research Centre, Faculty of Sciences and Technology, University of Coimbra, Coimbra, Portugal

# INTRODUCTION

The rapid industrialization and intensive agricultural activities over the last few decades have resulted in accumulation of various pollutants in the environment, which are distributed over wide areas by means of air and water. This has caused visible detrimental effects to the ecosystem and consequences to human health. Today, many soils throughout the world have undesirably high concentrations of heavy metals. These include lead (Pb), cadmium (Cd), zinc (Zn), mercury (Hg), arsenic (As), silver (Ag), chromium (Cr), copper (Cu), iron (Fe), and the platinum group elements. At low or background concentrations, heavy metals are not pollutants. They occur naturally in the environment due to their presence in bedrocks. Some heavy metals such as Zn and Cu are also essential micronutrients for living organisms. Therefore, the term *heavy metal pollution* refers to heavy metal levels that are abnormally high relative to normal background levels. All heavy metals at high concentration have strong toxic effects and are regarded as environmental pollutants.

Some heavy metals (like Fe, Zn, Ca and Mg) have been reported to be of bio-importance to man and their daily medicinal and dietary allowances have been recommended. However, some others (like As, Cd, Pb, and methylated forms of Hg) have been reported to have no known bio-importance in human biochemistry and physiology and consumption even at very low concentrations can be toxic [1]. Even for those that have bio-importance, dietary intakes have to be maintained at regulatory limits, as excesses result in poisoning or toxicity [2]. Although individual metals exhibit specific signs of their toxicity, the following have been reported as general signs associated with Cd, Pb, As, Hg, Zn, Cu and Al poisoning: gastrointestinal disorders, diarrhoea, stomatitis, tremor, ataxia, paralysis, vomiting and convulsion, depression, and pneumonia when volatile vapours and fumes are inhaled [3]. The nature of effects could be toxic (acute, chronic or sub-chronic), neurotoxic, carcinogenic, mutagenic or teratogenic.

Pb, Zn, Cu, Co, Mn, Fe, Cr and Cd have been found in the streams and rivers of the Americas, Europe, Asia, Africa and Australia [4-9]. In India, presence of heavy metals has been reported in the Brahmaputra [10]; the Kali and Hindon [11]; and more recently, in the Gomti [12]; the Cauvery [13]; and the Ganga [14].

The Yamuna (also Jamuna or Jumna) is the largest tributary of the Ganga in northern India, having the total length of about 1376 km. The source of Yamuna is Yamunotri in the Uttarakhand Himalaya, which is north of Haridwar in the Himalayan mountains. Yamuna river flows through the states of Uttarakhand, Delhi, Haryana and Uttar Pradesh and finally merges with river Ganges at a sacred spot known as Triveni Sangam in Allahabad. A number of prominent cities such as Delhi, Mathura and Agra lie on the bank of river Yamuna. Over 57 million people depend on the Yamuna waters. Just like the Ganges, the Yamuna too is highly venerated in Hinduism and worshipped as goddess Yamuna, throughout its course.

Due to high density population growth, rapid industrialization, today Yamuna is one of the most polluted rivers in the world, especially around New Delhi, where 15 drains discharge waste water into the river. The city dumps ~58% of its waste into it. When the river enters the city, it is already contaminated with 7500 coliform content per 100 ml. when it leaves the city, it carries with a dangerously high coliform content of 24 million per 100 ml. Even the ground water has been affected by leachates that pass down from the dumping sites. According to the Central Pollution Control Board (CPCB), 70% of the pollution in river is from untreated sewage and the remaining 30% is from industrial sources, agricultural run-off, garbage etc. The water quality of Yamuna River falls under the category "E" which makes it fit only for recreation and industrial cooling, completely ruling out the possibility for underwater life. Almost every year mass death of fishes is reported. Biological Oxygen Demand (BOD) load increased by 2.5 times between 1980 and 2005: from 117 tonnes per day in 1980 to 276 in 2005.

Although the government of India has spent nearly $500 million to clean up the river, the river continues to be polluted with garbage while most sewage treatment facilities are underfunded or malfunctioning. The Ministry of Environment and Forests (MoEF) of the Government of India (GOI) took measures to curb pollution in 12 towns of Haryana, 8 towns of Uttar Pradesh, and Delhi under an action plan (Yamuna Action Plan-YAP) which is being implemented since 1993 [15]. However in 2009, the Union government admitted the failure of the Ganga Action Plan (GAP) and the Yamuna Action Plan (YAP), saying that "rivers Ganga and Yamuna are no cleaner now than two decades ago" despite spending over Rs 1, 700 crore to control pollution [16].

In August 2009, Delhi Jal Board (DJB) initiated its plan for resuscitating a 22 km stretch of the Yamuna in Delhi by constructing interceptor sewers, at the cost of about Rs 1, 800 crore [17].

There are three main sources of pollution in the river, namely household and municipal disposal sites, agricultural run-off, and industrial effluents and run-off. Urban runoff and agricultural runoff are mainly non–point sources. The major sources of pollution from agriculture are fertilizers containing superabundant nutrients such as nitrogen and phosphorus, and heavy metals such as Cd, Cu, Pb and Zn. Water quality may also be altered by other factors, such as livestock manure, human waste, and atmospheric deposition. Atmospheric pollutants are often the largest source of waterborne metals. It is estimated that 70% of lead in water and over 50% of many of the other trace metals in the Great Lakes (USA) are derived from atmospheric transfer. In general, freshwater ecosystems have low natural background metal levels and therefore tend to be sensitive to even small additions of most trace metals. Heavy metal contamination of soils and water from industrial and traffic sources in urban environments has been studied in North America and Europe [18-22]. Agencies like the World Health Organization (WHO) and the United states Environment Protection Agency (USEPA) have set stringent standards for maximum permissible limits of heavy metals, but there is a paucity of detailed studies on heavy metal pollution and its remediation within industrial zones in developing countries. Yamuna outnumbers any other river in the number of industries on its bank. This is because it passes through many major industrial cities. About 22, 42, and 17 large and medium industrial units in the states of Haryana, Delhi, and Uttar Pradesh have been identified as polluting the river in the action plan area. In addition, the water in this river remains stagnant for almost 9 months in a year aggravating the situation.

According to the Agra District Industrial Centre officials, there were 226 iron foundries and about 340 metal casting units functioning in Agra in the decade of 1990-2000. Before the revised pollution control directives put the Agra diesel generator manufacturing industry off its track, the foundry industry of this town ranked among the country's largest assemblies of metal casting industrial units, generating business of over Rs 6, 000 crores. The ban on coking coal in the blast furnaces utilized by the foundry and metal-casting industry was a serious setback and the number of industrial units reduced drastically. In August 1999, the Supreme Court ordered the closure of 53 iron foundries and 107 other factories in Agra. In September 2010, it again ordered the closure of 212 of the 1, 715 small industries that had failed to disclose their toxic emission levels to the Uttar Pradesh Pollution Control Board (UPPCB). Another 299 were required to install pollution controlling devices, failing which they too would face closure. However, the ground realities are still nowhere near the reduced pollution levels targeted in Yamuna and its adjacent areas whether Agra or elsewhere, after it leaves the Himalayan foothills. The status quo, thus, ultimately leaves much to be desired.

Phytoremediation is an emerging technology that employs the use of green plants for the clean up of contaminated environment. It takes the advantage of the fact that a living plant acts as a solar-driven pump, which can extract and concentrate certain metals from the environment [23]. This remediation method maintains the biological properties and physical structure of the soil. The technique is environmentally friendly, cost-effective, visually unobtrusive, and offers the possibility of bio-recovery of the metals. In the case of heavy metal contamination in soil, phytoremediation techniques are narrowed down to *Phytoextraction,* where plants remove metals from the soil by concentrating them in their harvestable parts [24], and *Phytostabilization*, where plants reduce the mobility and bioavailability of pollutants by immobilization [25].

Phytoremediation is becoming possible because of the successful basic and applied research much of it conducted with the productive interdisciplinary cooperation of plant biologists, soil chemists, microbiologists and environmental engineers. Extensive progress has been made in characterizing and modifying the soil chemistry of the contaminated site to accelerate phytoremediation. The greatest progress in phytoremediation has been made with metals [26, 27]. Phytoremediation leaves the topsoil in usable condition and it is aesthetically pleasing. It requires minimal equipment and less energy inputs as plants do most of the work using solar energy. Thus, it is an eco-friendly process. The plants used can later be harvested, processed and disposed off in an environmentally sound manner. This technology has been receiving attention lately as an innovative, cost-effective alternative to the otherwise tedious and expensive methods in use which are not only a burden on the exchequer but also require efforts on recurring basis.

Phytoremediation employing indigenous species can be an ecologically viable option for sustainable and cost-effective management. Native plants often become adapted to locally elevated levels of metals in soil at contaminated sites, e.g. mines and industrial zones [28-30] and metal toxicity issues do not generally arise. Many native, well adapted plants have been investigated and even used for heavy metal bioindicatoring and phytoremedial purposes including lemongrass and other wild grasses, vetiver,*Sesbania, Avena, Crotalaria, Crinum asiaticum, Typha latifolia* and *Calotropis procera* etc. [31-35,28]. Native wild species are also important to remediate soils in context of the studied area due to a remark (April, 2006) of the Supreme Court prohibiting the cultivation of plants requiring fertilizers and pesticides along the Yamuna. In the light of this limitation, native wild species are a viable option since these do not require agronomic inputs.

Since the river Yamuna is the life line of Mathura and Agra, the existing pollution level has posed a serious threat not only to the environment but also to the human population. Adjacent areas are highly polluted and are a sink for a variety of chemicals including heavy metals. The present study was undertaken: (i) to get a comprehensive profile of eight metals in water and adjacent soils of the river Yamuna within Mathura, Agra and Bateshwar; (ii) to get a qualitative and quantitative estimate of the species present at test sites through phyto-sociological surveys; and (iii) to inventorize species with potential for phytoremediation present on sites by comparing with those previously reported by the authors as suitable in this context.

# CASE STUDY

Agra (27°10′N, 78°05′E, 169 msl), on the banks of the river Yamuna, is located in Uttar Pradesh in the north central part of India. It is roughly 200 km south-east of the national capital, New Delhi. Bounded by the Thar desert of Rajasthan on its south-east, west and north-west peripheries, it is a semi-arid area. The world renowned Mughal monument, the Taj Mahal is situated here. It is world renowned for its leather industry and marble handicrafts but it also boasts a cast iron and engineering goods industry. Mathura (27.28°N 77.41°E) is located approximately 60 km north of Agra and 145 km south-east of Delhi. According to Hindu scriptures Mathura is the birthplace of Lord Krishna. It is a fast expanding city with about half a million residents. Mathura oil refinery is one of the biggest oil refineries of Asia. Textile printing, dyeing and silver ornament manufacturing are major industries. Apart from these there are units manufacturing taps, household items, and cotton materials. Bateshwar (26.93°N 78.54°E) is a village on the banks of Yamuna about 120 km downstream from Agra. It is an important spiritual and cultural centre for Hindus.

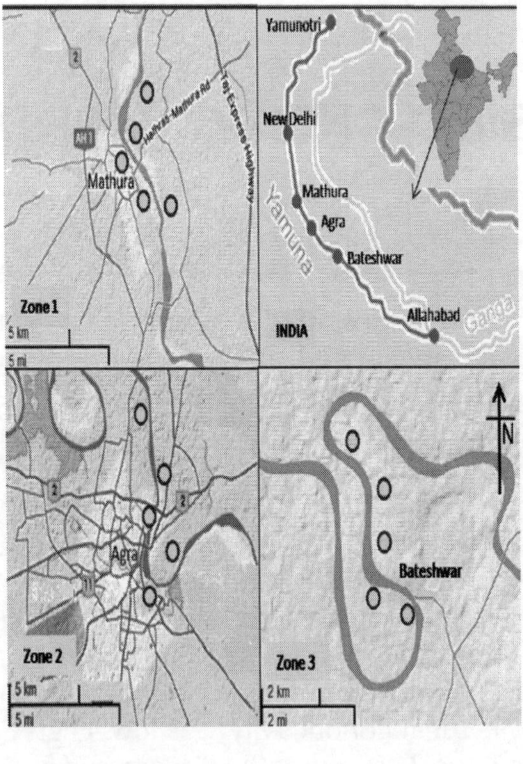


**Figure 1:** Map of the study area.

The study area was divided into three zones (Figure 1); all three along the course of Yamuna and covering two cities viz. Mathura (zone 1) and Agra (zone 2) and a large village i.e. Bateshwar (zone 3). The distance between zones 1 and 2 is 80 km and zones 2 and 3 is 120 km downstream. In all, a total distance of 200 km was covered along the course of river. In each zone, 5 sites were selected ~1 km apart. Five random soil samples were taken from 0-15 cm depth at each site.

A total of 75 soil samples (25 from each zone) were analyzed in order to obtain a complete profile. The same number of river water samples was collected from midstream at a depth of about 0.3 m. Soil from the botanical garden of St. John's college, Agra, was utilized as control.

The statistical significance of differences among mean metal content in water and soil was independently determined by one-way analysis of variance (ANOVA) followed by Fisher's LSD test. Pearson's coefficient for correlation of water and soil data was analyzed at a significance level of $P < 0.05$ and $P < 0.01$ with SPSS 16.0 statistics software.

# PHYSICO-CHEMICAL PROFILE

## Water

Physico-chemical properties of the water samples collected from the study zones are mentioned in Table 1. The pH values indicate neutral nature of river water acceptable as per BIS [36] and WHO [37] guidelines. A reading of 6.5 to 7.5 is considered neutral, suitable for general plant growth [38]. Conductance which reflects the status of major ions/inorganic pollution and is a measure of total dissolved solids and ionized species in the water, varies between 434 – 503 μmho/cm. Total dissolved solids were highest in zone 2. The hardness of water body is regulated largely by the levels of Ca and Mg salts. Other metals if present such as Fe, Al and Mn may also contribute to hardness. Most parameters were within their respective acceptable limits [36, 37]. Electrical conductivity was low. High COD, BOD and low DO in zones 1 and 2 are due to the discharge of huge amount of the untreated urban and industrial waste water/effluents indiscriminately. All three zones were faecally contaminated. Bacterial contamination ranged from19000—93000 coliform/100ml; the values are much higher than recommended values of 1coliform/100ml. Most of these coliforms were of faecal type due to gravity discharge of faecal wastes in adjacent areas along the river.

**Table 1:** Physico-chemical profile of water

| Parameters | **Zone** 1 | **Zone** 2 | **Zone** 3 | **Acceptable Limits [36, 37]** |
|---|---|---|---|---|
| pH (1:2.5) | 7.31 | 7.23 | 7.61 | 6.5-8.5 |
| Total Dissolved Solids (mg/l) | 266 | 314 | 245 | 500 |
| Conductivity (μmho/cm) | 462 | 503 | 434 | |
| N-$NO_3$ (mg/l) | 3.54 | 4.37 | 4.11 | 10 |
| N-$NH_3$ (mg/l) | 2.34 | 2.36 | 1.63 | 10 |
| Total hardness (mg/l) | 227 | 223 | 210 | 250 |
| Total alkalinity (mg/l) | 203 | 188 | 209 | |
| Chemical Oxygen Demand (mg/l) | 23.5 | 24.3 | 11.3 | |
| Chloride (mg/l) | 8.5 | 8 | 9 | 250 |
| Fluoride (mg/l) | 0.37 | 0.37 | 0.32 | |
| Dissolved Oxygen (mg/l) | 5.63 | - | 8.67 | >5 |
| Biological Oxygen Demand (mg/l) | 9.63 | 10.7 | 5.34 | |
| Sodium (mg/l) | 6.81 | 6.62 | 7.15 | |
| Potassium (mg/l) | 0.38 | 0.4 | 0.38 | |
| Ca hardness (mg/l) | 124 | 131 | 115 | |
| Mg hardness (mg/l) | 83 | 96.4 | 91.6 | |
| Faecal coliforms (MPN/100ml) | 86000 | 93000 | 19000 | |
| Streptococcus (MPN/100ml) | 64000 | 71000 | 17000 | |

## Soil

The soil of the study area is characterized by alluvium, which is an admixture of gravel, sand, silt and clay in various proportions deposited during the quaternary period. The area is a part of Indo-Gangetic alluvium of quaternary age and is made up of recent unconsolidated fluviatile formations comprising sand, silt, clay and *kankar* with occasional beds of gravel. The topsoil is coarse and angular sand with small clay fraction. The sub-soil is sandy throughout. The stabilized topsoil is reddish brown with sand and clay mixed. The minimum depth of topsoil layer is 60 cm.

Physico-chemical properties of soil samples are given in Table 2. The topsoil in the study area is sandy loam (sand 60-80%, silt 10-

24%, clay 8-16%). It has high exchangeable sodium percentage (ESP) values and moderate water retaining capacity. The sub-soil is sandy throughout. Soil pH ranged from neutral to alkaline. Zones 3, 2 and 1 were classified as very low, low and medium in organic matter, respectively.

**Table 2:** Physico-chemical profile of soils

| Zone | **pH (1:2.5)** | **Electrical Conductivity (dS/m) (1:2.5)** | **Organic Matter (%)** | **Avail. Phosphate (kg/ha)** | **Avail. Potash (kg/ha)** | **Avail. Nitrogen (kg/ha)** |
|---|---|---|---|---|---|---|
| 1 | 7.06-7.12 | 0.46-.50 | 1.44-1.53 | 108-115 | 236-298 | 53.4-60.8 |
| 2 | 6.24-6.81 | 0.33 0.38 | 0.8-1.2 | 131-140 | 65-94 | 50.1-55.2 |
| 3 | 7.43-7.6 | 0.44-0.47 | 0.5-0.72 | 50-65 | 143-178 | 75.2-87.8 |
| Control | 7.21 | 0.54 | 1.68 | 70.5 | 393 | 112.9 |

The electrical conductivity (EC) of soils ranged from 0.33-0.54 dS/m. Zone 1 and 2 soils fall in very high (>100 kg/ha), soils from zone 3 and control site in the high (50-100 kg/ha) phosphate availability bracket. Soils from zones 1, 3 and control displayed medium (130-330 kg/ha) potash levels while zone 2 was low (<130 kg/ha) in available potash. Nitrogen content in the soil samples ranged from 50.1 – 112.9 kg/ha.

# HEAVY METAL PROFILE

## Water

Concentrations of heavy metals in the water samples collected from different location have been summarized in Table 3. It is clearly evident from the table that heavy metals were consistently higher in zone 2 compared to zones 1 and 3. Cr content was markedly higher among the metals in zone 2 followed by zone 1. The concentration of heavy metals in water samples ranged from 0.018 – 0.095 mg Pb $l^{-1}$, 0.025

– 0.341 mg Cd $l^{-1}$, 0.47 – 1.76 mg Zn $l^{-1}$, 0.27 – 1.58 mg Cu $l^{-1}$, 0.001 – 0.005 mg Co $l^{-1}$, 0.80 – 9.37 mg Cr $l^{-1}$ and 0.078 – 0.32 mg Ni $l^{-1}$. As was not detected in any sample.

**Table 3:** Heavy metal content of water samples (mg $L^{-1}$)

| Zone | | Pb | Cd | Zn | Cu | Co | Cr | Ni |
|---|---|---|---|---|---|---|---|---|
| 1 | **Range** | 0.025-0.041 | 0.05-0.136 | 0.7-1.02 | 0.86-0.98 | 0.002-0.004 | 2.87-4.23 | 0.078-0.12 |
| | **Avg.** | 0.036a | 0.088a | 0.868a | 0.916a | 0.003a | 3.55a | 0.097a |
| | **SD** | 0.007 | 0.037 | 0.128 | 0.046 | 0.0005 | 0.540 | 0.016 |
| 2 | **Range** | 0.066-0.095 | 0.159-0.341 | 1.37-1.76 | 1.27-1.58 | 0.004-0.005 | 6.42-9.37 | 0.17-0.32 |
| | **Avg.** | 0.082b | 0.243b | 1.56b | 1.40b | 0.005b | 7.914b | 0.256b |
| | **SD** | 0.013 | 0.072 | 0.166 | 0.112 | 0.001 | 1.138 | 0.071 |
| 3 | **Range** | 0.018-0.028 | 0.025-0.03 | 0.47-0.61 | 0.27-0.33 | 0.001-0.002 | 0.8-0.97 | 0.009-0.015 |
| | **Avg.** | 0.023c | 0.028c | 0.54c | 0.3c | 0.001c | 0.864c | 0.01c |
| | **SD** | 0.004 | 0.004 | 0.058 | 0.042 | 0.0003 | 0.066 | 0.003 |
| | **F value** | * | * | * | * | * | * | * |
| **Permissible limits** | **WHO [37]** | 0.01 | 0.003 | 5 | 2 | - | - | - |
| | **USEPA [39]** | 0.015 | 0.005 | 5 | 1.3 | - | - | - |
| **World Average** | | 0.004 | 0.001 | 0.2 | 1.4 | - | - | - |

[i] - # As content below detection limit.

[ii] - F value:"*" statistically significant. Different letters in the same column denote significant statistical difference (P≤0.001) in mean metal contents in water samples from different zones

[iii] - SD- Standard deviation.

[iv] - WHO – World Health Organization.

[v] - USEPA – United States Environment Protection Agency

All the metals in water samples were positively ($P<0.01$) correlated with each other (Table 4). In other words, metal concentration trends were identical and increased simultaneously for Pb, Cd, Zn, Cu, Co, Cr, and Ni.

**Table 4:** Correlation coefficients: water heavy metal concentrations

| | Cd | Zn | Cu | Co | Cr | Ni |
|---|---|---|---|---|---|---|
| Pb | 0.970** | 0.985** | 0.875** | 0.924** | 0.960** | 0.962** |
| Cd | | 0.977** | 0.902** | 0.944** | 0.954** | 0.963** |
| Zn | | | 0.925** | 0.953** | 0.976** | 0.966** |
| Cu | | | | 0.963** | 0.940** | 0.899** |
| Co | | | | | 0.961** | 0.930** |
| Cr | | | | | | 0.947** |

[i] - ** Correlation is significant at the 0.01 level (two-tailed) (two-tailed; n=75)

Higher concentrations of metals in zone 2 (Figure 2) may be attributed to the discharge of industrial effluents from various sources including untreated sewage, municipal waste and agrochemical runoff from the nearby villages directly into the river. The concentrations of Co and Ni were found to be negligible at all sites. Due to the neutral to alkaline nature of river water, most of the heavy metals have precipitated and settled as carbonates, oxides, and hydroxide bearing sediments and elevated levels indicates higher exposure risks to the benthic biota of the river. Based on the WHO [37] and USEPA [39] drinking water standards (Table 3) the results in the present investigation show that Pb, Cd and Cr at all sites and Ni at most sites far exceeded the prescribed limits. Cu values from zone 2 were above the USEPA [39] threshold. One Way ANOVA and Fisher's LSD test indicate the difference in mean content of each metal among zones was highly significant statistically ($P \leq 0.001$).

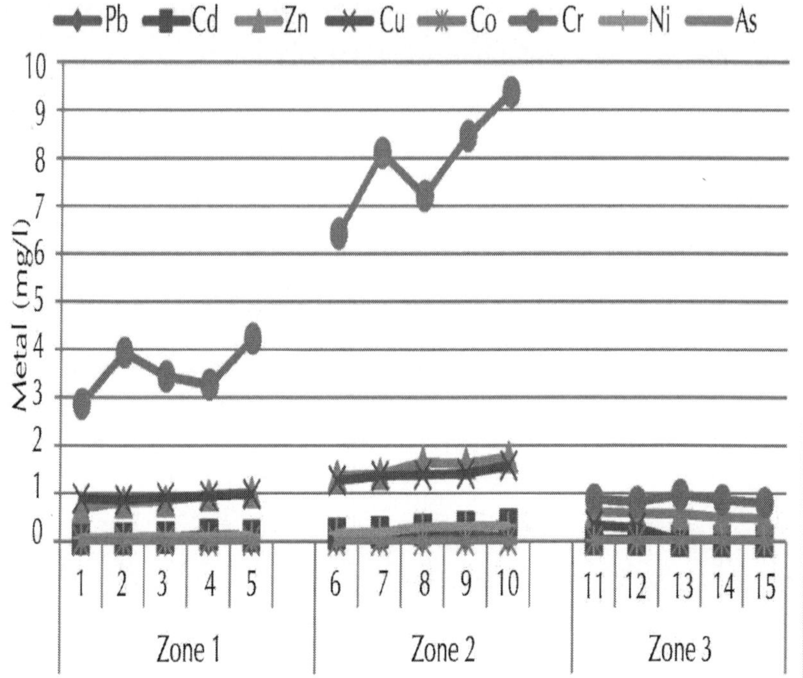


**Figure 2:** Average heavy metal content in water samples.

When compared with the metal profile of the rivers around the world (Table 5) the situation does not seem that desperate here, at least as far as heavy metal contamination is concerned. The picture, however, is quite different when we consider the WHO guidelines for drinking water and World average of trace elements in unpolluted rivers [56, 57], the concentration ranges of Pb and Cd were well above the international guidelines and acceptable concentrations for drinking water (Table 3). When compared to the world average of trace elements for unpolluted rivers, the river was considered polluted by Pb, Cd, Zn and Cu.

**Table 5:** Average heavy metal concentrations of rivers around the world (mg $L^{-1}$)

| Rivers | Pb | Cd | Zn | Cu | Co | Cr | Ni | References |
|---|---|---|---|---|---|---|---|---|
| **Yamuna river (present study)** | 0.018-0.095 | 0.05-0.341 | 0.47-1.76 | 0.27-1.58 | 0.001-0.005 | 0.8-9.37 | 0.009-0.32 | |
| **Cauvery river, India** | 13.35 | - | 47.51 | 4.57 | 8.25 | 1.01 | 4.53 | [13] |
| **Brahmaputra river, India** | - | - | 916 | 108 | 168 | 222 | 179 | [10] |
| **Ganga river, India** | 76.36 | 11.5 | 332.5 | 48.39 | - | 5.36 | 4.88 | [14] |
| **Gomti river, India** | 3.058 | - | 63.022 | - | - | 0.064 | 0.013 | [12] |
| **Challawa river, Nigeria** | 0.44 | - | 1.2 | 0.22 | - | 0.47 | - | [40] |
| **Mghogha river, Morocco** | 48.25 | 0.36 | 299.5 | 56.7 | - | 86.4 | 46.83 | [41] |
| **Sava river, Croatia** | 34 | 0.5 | 91 | 24 | - | - | - | [42] |
| **Pasig river, Philippines** | 70 | - | 530 | - | 160 | - | 21.2 | [43] |
| **Rhine river, Netherland** | 188.2 | 7.1 | 684.3 | 62.5 | | 6.4 | 33.7 | [44] |
| **Zhujiang, China** | 75.2 | - | 212 | 51 | 17.8 | 70.6 | 61.8 | [45] |
| **Almendares river, Cuba** | 93 | 2.5 | 262 | 158 | - | 90 | - | [46] |
| **Montevideo, Uruguay** | 44-128 | 1-1.6 | 174-491 | 58-135 | - | 79-253 | - | [47] |
| **Ribeira river, Brazil** | 767102 | 0.2-5.5 | 15-5090 | 60 | - | | - | [48] |

| | | | | | | | | |
|---|---|---|---|---|---|---|---|---|
| **Amazon river, Brazil** | 83 | - | 110 | 37.5 | - | 65 | 26.7 | [49] |
| **Danube river, Serbia and Montenegro** | 28.65 | 3.12 | 253.74 | 36.29 | - | 76.26 | 70.1 | [50] |
| **Msimbazi river, Tanzania** | - | 0.9 | 79 | 14 | - | 12 | 8.7 | [51] |
| **Brisbane River, Australia** | 20.1-81.9 | 1.9 | 40.8-144.0 | 31.1-30.2 | - | 14.2-54.3 | - | [52] |
| **Siahroud river, Iran** | 9.7 | 0.05 | 14.9 | - | - | 1.03 | - | [53] |
| **Gediz River, Turkey** | 1.3 | - | 2.6 | - | 1.6 | - | 4 | [54] |
| **Avg. shale value/ world avg.** | 20 | 0.3 | 95 | 45 | | 90 | 68 | [55] |

## Soil

Concentrations of heavy metals in the soil samples have been summarized in Table 6. Quantitatively the metals were observed in the sequence Pb > Zn > Cr > Ni > Cu > As > Cd > Co (Figure 3), though their thresholds for concern, mobility in soil and toxicity are different so this trend does not necessarily reflect the threat of individual metals. Pb and Zn were found in fairly higher concentrations at all the sampling locations. Generally, an overall linear increasing trend of metal contamination was noted from site 1, before the Yamuna enters the city of Mathura, to site 10 where the river leaves Agra. Thus, maximum values for all metals were observed in the samples pertaining to Agra. In the third zone metal concentrations were seen to decrease gradually. One-way ANOVA and Fisher's LSD test indicate that mean Pb and Co content was different at all sites ($P \leq 0.001$); while mean Cr, Cd, Cu, Ni, and As in zone 2 differed significantly from zone 1 and 3 ($P \leq 0.001$). The latter did not differ significantly among themselves. Mean Zn content in zone 1 differed significantly from zone 2 and 3 ($P \leq 0.05$). The difference between the latter was not significant statistically.

**Table 6:** Heavy metal content of soil samples (mg $kg^{-1}$)

| Zone | | **Pb** | **Cd** | **Zn** | **Cu** | **Co** | **Cr** | **Ni** | **As** |
|---|---|---|---|---|---|---|---|---|---|
| 1 | Range | 157-230 | 8.6-20.6 | 87.3-136 | 22.4-41.5 | 1.84-4.8 | 26.3-53.2 | 23.1-41.2 | 14.2-20.4 |
| | Avg. | 200a | 13.4a | 116a | 30.5a | 3.71a | 40.7a | 33.6a | 17.2a |
| | SD | 29.6 | 5.36 | 21.4 | 8.14 | 1.25 | 9.77 | 6.78 | 2.97 |
| 2 | Range | 241-285 | 17.8-25.2 | 129-222 | 50.4-64.2 | 5.91-15.2 | 76.3-104 | 57.8-71.3 | 22.2-28.6 |
| | Avg. | 261b | 20.3b | 173b | 57b | 9.6b | 86.7b | 63.5b | 25.2b |
| | SD | 16.1 | 3.13 | 37.6 | 5.93 | 3.80 | 10.5 | 5.44 | 3.10 |
| 3 | Range | 111-136 | 6.02-14.6 | 115-167 | 22.4-30.1 | 2.91-6.4 | 14.7-45.6 | 17.8-32.4 | 7.9-17.6 |
| | Avg. | 125c | 10.2a | 144b | 25.7a | 4.95c | 28.1a | 23.3a | 13.8a |
| | SD | 10.4 | 3.71 | 21.9 | 3.43 | 1.29 | 11.8 | 5.57 | 4.01 |
| **Control** | Range | 13.6-18.4 | 1.23-1.87 | 39.6-54.3 | 12.8-24.3 | 1.68-2.53 | 10.2-14.3 | 7.3-9.7 | 3.03-5.7 |
| | Avg. | 15.6 | 1.6 | 47.4 | 18 | 2.1 | 12 | 8.6 | 4.2 |
| | SD | 1.88 | 0.25 | 5.3 | 4.53 | 0.31 | 1.55 | 1 | 0.99 |
| | F value | * | * | * | * | * | * | * | * |
| **Suggested thresholds in soil [58]** | Industrial | 600 | 22 | 360 | 91 | - | 87 | 50 | 12 |
| | Residential | 140 | 10 | 200 | 63 | - | 64 | 50 | 12 |

| | | | | | | | | | |
|---|---|---|---|---|---|---|---|---|---|
| **Suggested thresholds in soil [59]** | Background | - | - | 140 | - | - | 100 | 35 | 29 |
| | Intervention | - | - | 720 | - | - | 380 | 210 | 55 |
| **Threshold values [60]** | Class I | 35 | 0.2 | 100 | 35 | - | 90 | 40 | 15 |
| | Class II | 250 | 0.3 | 200 | 50 | - | 150 | 60 | 30 |
| | Class III | 500 | 1.0 | 500 | 400 | - | 300 | 200 | 40 |

[i] - F value:'*' statistically significant. Different letters in the same column denote significant statistical difference (P≤0.05) in mean metal contents in soil samples from different zones.

[ii] - SD- Standard deviation

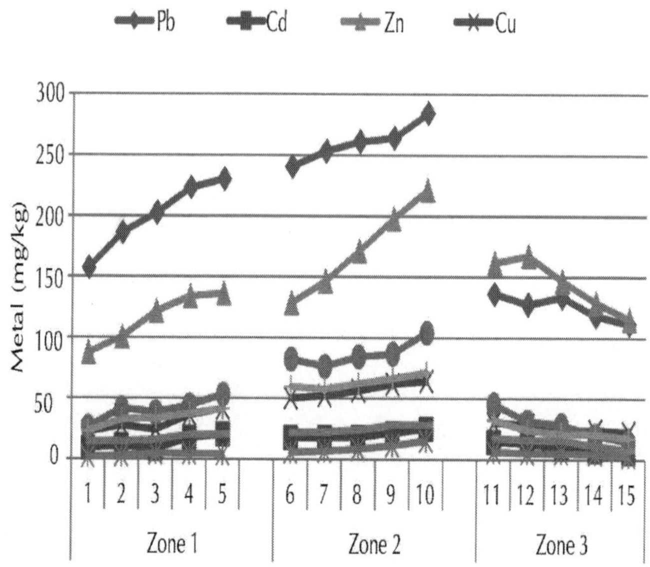


**Figure 3:** Average heavy metal content in soil samples.

All the metals in soils were positively (P<0.01) correlated with each other (Table 7). Significant negative correlation was observed between metal concentrations and soil pH (P<0.01). The same was observed in the case of Zn and Co with Organic matter. Phosphate is able to increase water-soluble lead forms from contaminated soils by 56.8– 100% [61]. This is clearly shown by the phosphate values (Table 2) obtained for different samples with maximum in zone 2 followed by zone 1 which probably led to higher Pb values in zones 1 and 2 (Table 6). Fertilizers contain from trace to several ppm of Pb, Zn, Cu, Mg [62, 63]. High $P_2O_5$-blended fertilizers and the pure phosphates, contain significant concentrations of several elements of potential environmental or agronomic concern [62, 64].

**Table 7:** Correlation coefficients: soil heavy metal concentrations

| | Cd | Zn | Cu | Co | Cr | Ni | As | OM | pH |
|---|---|---|---|---|---|---|---|---|---|
| **Pb** | 0.821** | 0.479** | 0.889** | 0.629** | 0.909** | 0.933** | 0.894** | 0.426** | -0.802** |
| **Cd** | | 0.701** | 0.886** | 0.724** | 0.862** | 0.848** | 0.906** | 0.0876 | -0.621** |
| **Zn** | | | 0.723** | 0.915** | 0.689** | 0.653** | 0.744** | -0.493** | -0.371** |
| **Cu** | | | | 0.805** | 0.966** | 0.972** | 0.932** | 0.0260 | -0.773** |
| **Co** | | | | | 0.809** | 0.784** | 0.806** | -0.304** | -0.529** |
| **Cr** | | | | | | 0.991** | 0.944** | 0.0816 | -0.834** |
| **Ni** | | | | | | | 0.936** | 0.138 | -0.837** |
| **As** | | | | | | | | 0.106 | -0.739** |
| **OM** | | | | | | | | | -0.242* |

[i] - * Correlation is significant at the 0.05 level (two-tailed; n=75)

[ii] - ** Correlation is significant at the 0.01 level (two-tailed)

Agra is the fourth most populated city in Uttar Pradesh, India. With a population of 1.7 million (2011 census) it generates about 700 tonnes of solid wastes every day. It is also a major cause for adding contamination to soil and groundwater. Solid waste is also discharged from 200 hospitals and nursing homes along with 168 foundries, 52 tanneries, 300 shoe industries, 200 petha (a local sweet) manufacturing units, 50 dairies, 56 electroplating units, 15 silver vibrators and 15 galvanizing units. Significantly higher amount of metal pollution in the samples from the city (sites 6-10) is obviously due to untreated domestic/wastewater, sewage and industrial effluent discharged at these sites throughout the year. The increasing contamination as one proceeds downstream mirrors the extent of damage caused to the pedosphere.

Mean concentrations of heavy metals in soils at the sites studied were compared with threshold values of soil suggested by the Canadian Environmental Quality Guidelines [58]. It was observed that As (sites 1-13) and Ni (sites 6-10) crossed their respective industrial thresholds while the other metals (Pb, Zn and Cu) are well within it. Mean concentrations of As at sites 4-10 were approximately twice the thresholds suggested. Cd and Cr levels were above their thresholds only at site 10. However, the situation is drastically different in the perspective of the residential limits where in addition to these, the thresholds are exceeded even by Pb, Cd (10 sites each), Cr (5 sites) and also Zn and Cu at one site.

On comparing metal concentrations with the values suggested for soil remediation by VROM, Netherlands [59], values of Zn (sites 7-13), Ni (sites 4-10) and Cr (site 10) were above the background values but below the intervention level. It is significant to note that in studies similar to the present one, the degree of contamination and the resulting 'hazard indices' for soils may vary when different thresholds, existing in only a few countries, are considered [65]. To increase the reliability of risk estimation due to contaminants, global consensus on such thresholds is urgently needed.

The concentrations of As are usually low, less than 6 ppm, for geological and soil environment [64]. It is estimated that about 60% As in the environment is from anthropogenic sources including As-based pesticides, fertilizers, and wastes from mines, smelter and tannery industries [66]. The relatively high values of As in the samples seem to be directly related to the discharge of domestic and industrial effluent as well as use of phosphate fertilizers, pesticides used in the agricultural activities in the region.

Highly significant positive correlation ($P<0.01$) was observed between soil and water content of Pb, Cd, Cu, Co, Cr and Ni. The results also indicate that metal concentrations in soil were higher than those in the water. This distribution pattern of heavy metals between the water phase and soil is expected as most heavy metal speciation studies have reported a similar pattern of distribution both in sea water as well as in lakes [67-69].

Several authors have pointed out the need for a better knowledge of urban soils [18, 70]. In the past few years, studies on urban soils in many cities have been carried out around the world. Some examples are Spanish [19, 71] and Italian cities [21, 72]. Other examples for European cities are Aberdeen [73], Athens [74], Oslo [22] and Belgrade [18]. The mean heavy metal contents for all zones are compared in Table 8 to those of some cities around the world. The differences concerning population, living habits, industrial activities, etc., cause significant differences in the metal contamination profile. Compared to average concentrations in urban soils in the world, the mean concentrations of Pb and Cu are up to 2—4 times higher in some cases but still less than London, Naples and Palermo. In the case of Cd, it is many times higher than Kattedan (India). Zn and Cr contents do not differ much; still they are less than those of Naples and Madrid. Ni content is more than almost all European cities, but less than Kattedan and Firozabad in India. Co values are less than those reported from other industrial regions of India. As content is less than that of Firozabad.

**Table 8:** Average heavy metal concentrations in urban soils from different cities across the world (mg $kg^{-1}$)

| City | | Pb | Cd | Zn | Cu | Co | Cr | Ni | As | Reference |
|---|---|---|---|---|---|---|---|---|---|---|
| **London** | | 294 | - | 183 | 73 | - | - | - | - | [75] |
| **Madrid** | | 161 | - | 210 | 72 | - | 75 | 14 | - | [76] |
| **Rostock** | | 83 | - | 100 | 35 | - | 48 | 30 | - | [77] |
| **Sevilla** | | 161 | - | 107 | 64.6 | - | 42.8 | 23.5 | - | [19] |
| **Belgrade** | | 53.2 | - | 129.1 | 29 | - | 33.2 | 67.4 | - | [18] |
| **Palermo** | | 253 | - | 151 | 77 | - | 39 | 19.1 | - | [72] |
| **Naples** | | 262 | - | 251 | 11 | - | 74 | - | - | [21] |
| **Nanjing** | | 107.3 | - | 162.6 | 66.1 | - | 84.7 | - | - | [78] |
| **Hong Kong** | | 93.4 | - | 168 | 24.8 | - | n.a. | - | - | [79] |
| **Kattedan** | | 195-6241 | 0.08-0.16 | 130-3191 | 72-1450 | 12-36 | 77-586 | 63-494 | 0.10-0.21 | [80] |
| **Firozabad** | | 35.5-781 | 3.64-107 | 76.4-1247 | 22.4-300 | 10.9-63.7 | 19.1-158 | 23-218 | 9.25-204 | [29] |
| **Present Study** | **Zone 1** | 200 | 13.4 | 116 | 30.4 | 3.70 | 40.7 | 33.6 | 17.2 | |
| | **Zone 2** | 261 | 20.3 | 173 | 57.0 | 9.60 | 86.7 | 63.5 | 25.2 | |
| | **Zone 3** | 125 | 10.2 | 144 | 25.7 | 4.96 | 28.1 | 23.3 | 13.8 | |

It is encouraging to note that the mean concentrations of individual metals are below those reported from other industrial hubs within India i.e. Kattedan (Andhra Pradesh) [except Cd and As] and Firozabad (Uttar Pradesh). Kattedan Industrial Development Area (KIDA) is a major industrial area of Andhra Pradesh and houses 400–500 industries, including 150 large scale industries and 300 small-scale industries. Major sources of metals pollution are battery, electrode, oil refining, metal plating, textile, pharmaceutical, chemical paints, rubber, petrochemicals, glass, therapeutics, and Pb extraction facilities [81]. This is also one of the contaminated areas identified by the Central Pollution Control Board (CPCB) in New Delhi, and referred to as an ecological disaster area [81]. Firozabad is the hub of the Indian Glass industry.

# ASSESSMENT OF HEAVY METAL CONTAMINATION IN SOIL

Assessment of soil contamination was performed by the contamination index ($P_i$) and integrated contamination index ($P_c$) as expressed by fuzzy functions [82, 29, 28]. Class I criteria [60] could be used as no-polluted threshold; Class II as lowly polluted threshold value; and while Class III as highly polluted threshold value. $P_i$ values ≤ 1 indicate no contamination; $1 \leq P_i \leq 2$ indicates low contamination; $2 \leq P_i \leq 3$ indicates moderate contamination; while $P_i > 3$ indicates high contamination.

Individual elements displayed remarkably different patterns of accumulation in soils. Furthermore, observed differences in the magnitude of accumulation suggest that the relative contribution of the individual elements to total heavy metal contamination varies. Figure 4 shows the proportions of contamination levels (from $P_i$ values) in the soil samples from all the sites studied. Except for 76% samples from zone 2, which showed moderate Pb contamination, the rest exhibited low contamination zone as did all samples from zones 1 and 3. In case of Cd, all samples were in the high contamination zone. For Zn, 24% samples from zone 2 were moderately contaminated while 72%, 76%, and 100% samples from zones 1, 2, and 3, respectively were in the low contamination range. For Cu, 88% samples from zone 2 were moderately contaminated while 36% and 12% samples, from zones 1 and 2, respectively were in the low contamination range. All samples from zone 3 indicated no contamination. Except for zone 2 (20% samples) in the low contamination zone, the remaining samples did not indicate Cr contamination. For Ni, 12% and 72% samples from zones 1 and 2, respectively were moderately contaminated while 16% and 22% samples from, respectively were in the low contamination range. All samples from zone 3 indicated no contamination. In the case of As, 64%, 100%, and 60% samples from zones 1, 2, and 3 were in the low contamination range.

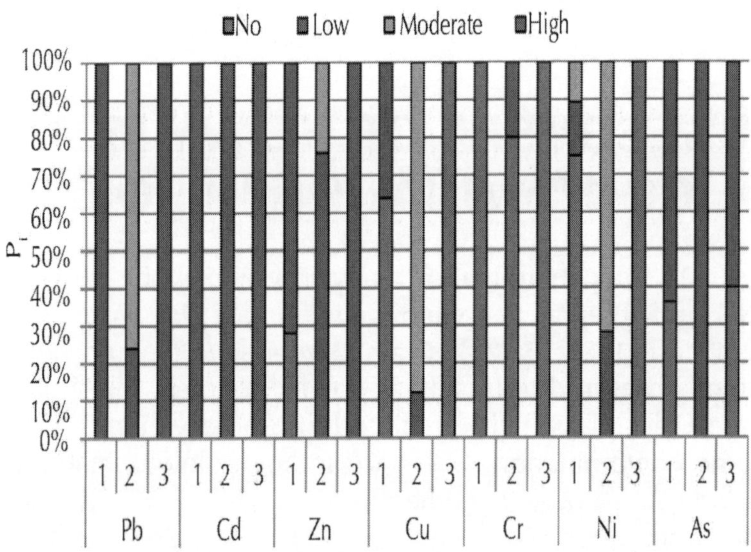


**Figure 4:** Contamination indices (Pi) of heavy metals in soil samples.

Thus, zone 1 was found to be lowly contaminated with Pb, Zn, Cu, Ni and As but highly contaminated with Cd. Zone 2 exhibited low to moderate contamination of Pb, Zn, Cu, Ni; low Cr and As contamination; and high Cd contamination. Zone 3 was lowly polluted with Pb and Zn. As contamination ranged from none to low. No Cu, Cr and Ni contamination was observed. These results agree with the findings regarding metal contamination of soil due to the glass industry at Firozabad, India [29]. Of the nine elements studied, Zn, Cd, and As showed a greater accumulation in all soils, whereas, accumulation of Ni and Cu was high in limited samples.

Integrated Contamination Indices ($P_c$) were calculated for all soils to assess the extent of heavy metal contamination at the sites. $P_c$ is defined as the summation of the difference between the contamination index for each metal and 1 (one). It is categorized under the following heads: $P_c \leq 0$ no contamination; $0 \leq P_c \leq 7$ low contamination; $7 \leq P_c \leq 21$ moderate contamination; $P_c > 21$ high contamination. Threshold values for Co could not be obtained hence this metals was excluded in the calculation. A clear ascending trend is visible in the $P_c$ values for all sites (Figure 5). $P_c$ values generally show a moderate to high contamination at studied sites. The $P_c$ indices indicate that 45% sampling locations fall in the moderate contamination while 55% of the samples fall in the high contamination range. All the samples from zone 2 fell under the high contamination category. While in zone 1, 60% samples come under moderate and 40% under high contamination level category. In zone 3, 76% and 24% samples were in the moderate and high contamination range, respectively.

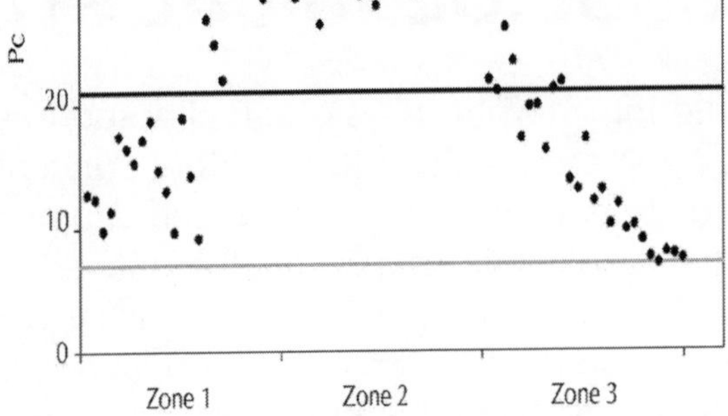


**Figure 5:** Integrated contamination indices (Pc) of soil samples. Black and gray lines are the upper threshold values of moderate and low contamination, respectively.

The effect of the glass industry on urban soil metal characterization was assessed at 25 test sites at Firozabad, India [29]. The area is characterized by little or no monitoring of industrial processes, usage and disposal of hazardous chemicals. A comprehensive profile of Zn, Mn, Co, Cd, Pb, Cr, Ni, Cu and As contamination was obtained. Zn, Cd, and As showed a greater accumulation in all soils, whereas, accumulation of Ni and Cu was high in limited samples. Integrated contamination indices ($P_c$) indicate that 60% of the sites were in the high contamination range and 28% were in the moderate contamination range with just 12% sites on the border of the moderate to low contamination range. [83] assessed the impact of both landuse and soil textures on Cd, Zn, Pb and Cu based on samples collected from the major landuse/landcover pattern of Dutch forests and aerable soils drawn from six different sites. Metal content in agricultural and industrial soil is found to be higher than the forest soil.

The fact that no $P_c$ value in the present investigation fell within the low contamination range was not surprising, given the fact that the study was being carried out in an area which has already been contaminated with metals, but moderate to high indices in zone 1 and 2 are alarming because these include heavily populated areas. The local populace is, thus, exposed to wide range of historically well established toxins and even carcinogens. The situation is surely compounded by vehicular pollution at urban sites (1-10). Vehicular emissions are a significant source of many pollutants [21, 84].

# PHYTOSOCIOLOGICAL STUDIES

Plants show differing morpho-physiological responses to soil metal contamination. Most are sensitive to very low concentrations; others have developed tolerance, and a reduced number accumulate metals. The latter capacity has practically opened up the way to phytoextraction. Hyperaccumulation is an unusual occurrence, seen in a narrow range of species which often grow in metal-rich soils. The following thresholds for metal hyperaccumulation in shoots, without evident symptoms of toxicity, have been suggested [85]: 100 mg $kg^{-1}$ for Cd, 1, 000 mg $kg^{-1}$ for Ni, Cu, Co and Pb, and 10, 000 mg $kg^{-1}$ for Zn and Mn. Known hyperaccumulators are generally minor vegetation components in most European and North American habitats. Currently, more than 400 hyperaccumulator species are known, belonging to 45 different botanical families, among which the most frequent are Brassicaceae and Fabaceae [86].

Lack of information on the agricultural management of hyperaccumulators, together with slow-growing and poor shoot and root growth, increase the difficulties in the practical application of these species in remediation projects [87]. Hence, the potential for any plant species to remediate successfully heavy metal contaminated sites depends on all of the following prerequisite factors: a) the amount of metals that can be accumulated by the candidate plant, b) the growth rate of the plant in question, and c) the planting density [88]. The growth rate of a plant in a chemically contaminated soil is important from the perspective of biomass. Parameters like basal area, canopy, abundance, dominance of species can help obtain a more rounded picture in the case of mixed planting or natural flora at a contaminated site. The rate of metal removal from the soils can be calculated if information on the above mentioned parameters is available. In addition, versatility of the candidate plant to tolerate and at the same time accumulate multiple metal contaminants and/or metal-organic mixtures would be an asset for any phytoremediation system.

The choice of plant species is thus, an important task in any phytoremediation based technique. Decontaminating a site in a reasonable number of harvests requires plants that are both high yielders of biomass and good metal accumulators by dry weight. It has been demonstrated [89, 90] that, wild native plants may be better phytoremediators for waste lands than the known metal bioaccumulators like *Thlaspi caerulescens* and *Alyssum bertolonii* because the latter are slow growing with shallow root systems and low biomass. Also, the technology for their large-scale cultivation is not fully developed; therefore, their use is rather limited [91].

If soil at contaminated sites, e.g. mines, industrial zones is naturally high in a particular metal, native plants will often become adapted over time to the locally elevated levels [28-30] and metal toxicity issues do not generally arise. Successful establishment and colonization of several pioneer plant species tolerant to Pb/Zn mine spoils has also been demonstrated with tolerant plants including *Phragmites australis, Vetiveria zizanioides,* and *Sesbania rostrata* [31, 92]. Many native, well adapted plants have been investigated and even used for heavy metal bioindicatoring and phytoremedial purposes including lemongrass and other wild grasses, vetiver, *Sesbania, Avena, Crotalaria, Crinum asiaticum, Typha latifolia* and *Calotropis procera* etc. [31-35]. Phytoremediation employing indigenous species can be an ecologically viable option for sustainable and cost-effective management.

An important component of any ecosystem is the species it contains. Species also serve as good indicators of the ecological condition of a system [93]. Ecological surveys are necessary for an adequate characterization of a plant community and also to know the diversity and dispersion status of species in the area. Phytosociology aims to characterize and classify plant communities in terms of composition and structure.

At all sampling sites within a zone, ecological indices [relative frequency, relative density, relative dominance and importance value index (IVI)] were estimated, by using a 1 $m^2$ quadrat. Sampling was done randomly at 10 spots at each site within a zone. The data were compiled and analysed according to some workers [94-96].

Relative density is the proportion of density of a species (plants/unit area) to that of the stand as a whole. The dispersion of species in relation to that of all the species is termed as relative frequency of a species. Relative dominance is the proportion of the basal area of a species to the sum of the basal area of all species present. Basal area refers to area covered by the plant's stem and leaves one inch above the ground surface. The overall picture of ecological importance of a species in relation to the community structure can be obtained by adding the values of the above three parameters [97].

A total of 22 weed species were recorded from the sites (Table 9). Most of the weeds recorded are herbs except *Calotropis procera* and *Datura stramonium* which are shrubby in nature. Two grasses i.e.*P. annua* and *C. dactylon* were observed. The phytosociological parameters obtained from the sites clearly indicate that there are naturally occurring plant species which have the capacity to tolerate the heavy metal content of the soils. The floral composition of the three zones varied qualitatively and quantitatively. Most species were seen to grow vigorously. Relative frequency, relative density, relative dominace and IVI indicate that *Calotropis procera, Parthenium hysterophorus, Chenopodium murale, Croton bonplandianum, Rumex dentatus, Amaranthus spinosus, Datura stramonium* and *Withania somnifera* were the most abundant weeds.

All of these species have been reported as potential phytoremediators in earlier studies. It is important to note that floral diversity decreased with increasing contamination profile of the sites. Maximum species (20) were observed in zone 3, followed by zones 1 and 2.

**Table 9:** Phytosociological parameters of flora at test sites

| Zone | **Plants** | **Relative frequency** | **Relative density** | **Relative dominance** | **IVI** |
|---|---|---|---|---|---|
| 1 | *Amaranthus spinosus* | 8.42 | 10.55 | 0.15 | 19.12 |
| | *Rumex dentatus* | 6.32 | 6.00 | 0.07 | 12.38 |
| | *Calotropis procera* | 8.42 | 10.31 | 0.18 | 18.92 |
| | *Croton bonplandianum* | 6.32 | 6.00 | 0.23 | 12.54 |
| | *Chenopodium murale* | 6.32 | 6.24 | 0.01 | 12.56 |
| | *Datura stramonium* | 7.37 | 9.83 | 0.16 | 17.37 |
| | *Stellaria media* | 5.26 | 4.32 | 0.00 | 9.58 |
| | *Withania somnifera* | 8.42 | 9.59 | 0.13 | 18.15 |
| | *Heliotropium ellipticum* | 5.26 | 3.12 | 0.00 | 8.38 |
| | *Achyranthes aspera* | 6.32 | 5.76 | 0.01 | 12.08 |
| | *Parthenium hysterophorus* | 7.37 | 6.95 | 0.01 | 14.33 |
| | *Amaranthus alba* | 5.26 | 4.56 | 0.01 | 9.83 |
| | *Boerhaavia diffusa* | 3.16 | 2.64 | 0.01 | 5.80 |
| | *Euphorbia hirta* | 6.32 | 6.00 | 0.00 | 12.31 |
| | *Sida longifolia* | 4.21 | 4.56 | 0.00 | 8.77 |
| | *Gnaphalium luteo-album* | 5.26 | 3.60 | 0.02 | 8.88 |
| 2 | *Parthenium hysterophorus* | 11.59 | 16.32 | 0.22 | 28.13 |
| | *Abutilon indicum* | 7.25 | 4.75 | 0.06 | 12.06 |
| | *Calotropis procera* | 11.59 | 15.13 | 0.20 | 26.93 |
| | *Croton bonplandianum* | 7.25 | 8.90 | 0.14 | 16.28 |
| | *Amaranthus spinosus* | 4.35 | 2.08 | 0.00 | 6.43 |
| | *Rumex dentatus* | 11.59 | 13.65 | 0.15 | 25.39 |
| | *Withania somnifera* | 8.70 | 8.61 | 0.10 | 17.40 |
| | *Cynodon dactylon* | 7.25 | 3.56 | 0.00 | 10.81 |
| | *Chenopodium murale* | 7.25 | 9.50 | 0.11 | 16.85 |
| | *Achyranthes aspera* | 5.80 | 4.75 | 0.02 | 10.57 |
| | *Sida longifolia* | 7.25 | 2.97 | 0.00 | 10.21 |
| | *Gnaphalium luteo-album* | 5.80 | 4.15 | 0.00 | 9.95 |
| | *Boerhaavia diffusa* | 4.35 | 5.64 | 0.00 | 9.99 |

| | | | | | |
|---|---|---|---|---|---|
| 3 | *Parthenium hysterophorus* | 7.14 | 9.65 | 0.17 | 16.97 |
| | *Abutilon indicum* | 4.76 | 3.46 | 0.01 | 8.23 |
| | *Calotropis procera* | 7.14 | 12.57 | 0.18 | 19.89 |
| | *Croton bonplandianum* | 5.56 | 5.65 | 0.17 | 11.37 |
| | *Cynodon dactylon* | 4.76 | 2.91 | 0.06 | 7.74 |
| | *Chenopodium murale* | 6.35 | 12.39 | 0.14 | 18.88 |
| | *Poa annua* | 4.76 | 3.28 | 0.01 | 8.05 |
| | *Rumex dentatus* | 3.97 | 1.64 | 0.01 | 5.62 |
| | *Barleria diffusa* | 4.76 | 4.55 | 0.00 | 9.32 |
| | *Achyranthes aspera* | 5.56 | 4.19 | 0.01 | 9.76 |
| | *Sida longifolia* | 4.76 | 3.64 | 0.00 | 8.41 |
| | *Withania somnifera* | 4.76 | 3.83 | 0.09 | 8.68 |
| | *Boerahvia diffusa* | 4.76 | 2.55 | 0.00 | 7.31 |
| | *Sida cordifolia* | 4.76 | 3.28 | 0.00 | 8.04 |
| | *Amaranthus spinosus* | 7.14 | 8.56 | 0.10 | 15.80 |
| | *Gnaphalium luteo-album* | 4.76 | 3.83 | 0.03 | 8.62 |
| | *Euphorbia hirta* | 3.17 | 3.46 | 0.00 | 6.64 |
| | *Ageratum conyzoides* | 3.17 | 2.37 | 0.00 | 5.55 |
| | *Datura stramonium* | 4.76 | 4.19 | 0.00 | 8.95 |
| | *Tridex procumbens* | 3.17 | 4.01 | 0.00 | 7.18 |

[i] - IVI- Importance Value Index

*C. procera* has been demonstrated as a potential phytoremediator species. The shrub showed good accumulation of metals and is a potential phytoextractor for As and Zn as well as a promising phytostabiliser for Pb, Cd, Cu and Mn [28, 29, 35]. *C. procera* was observed to have a high degree of sociability i.e. relative frequency, relative density, relative dominance and IVI. *P. hysterophorus* was also important in this context and was most dominant in zone 2. This species has been identified for As phytoextraction along with *A. spinosus, C. bonplandianum,* and *D. stramonium* [28]. The latter two have also been indicated for phytoextraction—*C. bonplandianum* for Mn and *D. stramonium* for Mn, Cr, and Cu—together with *R. dentatus* for Pb. Another species with high IVI, *C. murale* has been suggested for Zn, Cd, Pb and Cu phytoextraction [28, 29]. Among the less dominant species, *Tridex procumbens* and *Euphorbia hirrta* have also been reported as promising tools for phytoextraction of Mn and As, respectively [28]. *E. hirrta* and *D. stramonium* were not found in zone 2.

*Poa annua* has been identified as a phytostabilizer for Mn, Cd, and As and phytoextractor for Cu and Pb. Cu concentrations up to 742.06 mg $kg^{-1}$ dry weight have been reported in *P. annua* shoots [28,29]. *Poa annua* was observed only at sites in zone 1.

Other species found at the sites have also been indicated for further studies following initial field surveys. *Gnaphalium luteo-album* (Mn and As); *Withania somnifera* (Cu); and *Heliotropium ellipticum*(As) have shown promise as phytostabilisers for these metals and metal combinations [28, 29].

# DISCUSSION

The occurrence as well as concentrations of heavy metals like Pb, Zn, Cu, Co, Mn, Fe, Cr and Cd in streams and rivers all over the world is increasing. In the present case study, heavy metal contamination was consistently higher in city of Agra, which may be attributed to the heavy industrialization combined with agricultural and urban runoff. The situation is made worse by atmospheric deposition, again attributable to industrial and vehicular pollution. In general, freshwater ecosystems have low natural background metal levels and therefore tend to be sensitive to even small additions of most trace metals. The river water far exceeded the limits of metals prescribed by WHO and USEPA for drinking water standards and Pb, Cd, and Cr content at all sites and Ni at most sites exceeded the prescribed limits. In a heavily populated country like India where a sizeable portion of the population is illiterate and resides in slums/poorly planned neighbourhoods without proper sanitation and drainage, day-to-day activities also contribute to the overall pollution load. Provision of suitable alternatives along with proper education and awareness is integral to the minimization of this problem at the source. Apart from taking measures like effluent treatment before it enters the river and subsequent treatment of river water at the most polluted sites, a steady flow of water is to be ensured throughout the year, by way of channelizing the river with canals at crucial points. Such measures can address this problem to a substantial extent. Expenditure of more than US$ 500 million without much success appears to be an unjustified proposition.

Phytoremediation has been receiving attention lately as an innovative, cost-effective alternative to the otherwise tedious and expensive methods in use, which are not only a burden on the exchequer but also, require efforts on a recurring basis. Lack of information on the agricultural management of hyperaccumulator species, together with their poor biomass and root proliferation, increases the difficulties in their practical application. It has been amply demonstrated that wild native plants may be better phytoremediating tools. These species can be an ecologically viable option for sustainable and cost-effective management especially in scenarios where expertise, technical expertise and/or funding is a limiting factor. Ecological surveys are necessary for adequate characterization of a plant community and subsequent identification of prospective candidates for phytoremedial strategies since metal toxicity issues generally do not arise in plants already established on contaminated soils. Allowing native species to remediate site is an attractive proposition since these species do not require frequent irrigation, fertilizers, and pesticide treatments, while simultaneously a plant community comparable to that existing in the vicinity can be established. The outcome is, thus site remediation, ecological restoration and addition in aesthetic value. This is also in concurrence with the ruling (2006) of the Hon'ble Supreme Court of India prohibiting cultivation of plants requiring fertilizers and pesticides along the river Yamuna. Using these perennial phytoremedial candidates without any special needs holds much promise in this context. In addition, versatility of the candidate plant to tolerate and at the same time accumulate multiple metal contaminants and/or metal-organic mixtures would be an asset for any phytoremediation strategy.

## ACKNOWLEDGEMENTS

Financial support from University Grants Commission [F. no. 35-47/2008(SR)] is gratefully acknowledged. This study was partially supported by the European Fund for Economic and Regional Development (FEDER) through the Program Operational Factors of Competitiveness (COMPETE) and National Funds through the Portuguese Foundation for Science and Technology (PEST-C/MAR/UI 0284/2011, FCOMP 01 0124 FEDER 022689.

# REFERENCES


1. Nolan K. Copper Toxicity Syndrome. Journal of Orthomolecular Psychiatry 2003;12(4) 270–282.
2. Marschner H. Mineral Nutrition of Higher Plants. London: Academic Press Limited; 1995.
3. McCluggage D. Heavy Metal Poisoning, NCS Magazine. Columbus, USA: The Bird Hospital; 1991. http://www.cockatiels.org/articles/ Diseases/metals.html.
4. Katz BG, Bricker OP, Kennedy MM. Geochemical mass balance relationships for selected ions in precipitation and stream water, Catoctin Mountains, Maryland. American Journal of Science 1985;285 931–962.
5. Houba C, Remade J, Dubois D, Thorez J. Factors affecting the concentrations of cadmium, zinc, copper and lead in the sediments of the Vesdre river. Water Research 1983;17 1281–1286.
6. Zhang J, Huang WH. Dissolved trace metals in the Huanghe: The most turbid large river in the world. Water Research 1993;27 1–8.
7. Pizarro J, Vila I, Manuel C. Chemical compositions of Chilean rivers. Verth-Int. Ver. Theory of predictive Limnology 1998;26 948–951.
8. Alexandrine E, Jean-Marie M, Thevenot DR. Urban runoff impact on particulate metal concentrations in the river Seine. Water, Air, and Soil Pollution 1998;108 83–105.
9. Pistelok F, Galas W. Zinc pollution of the Przemsza river and its tributaries. Polish Journal of Environmental Studies 1999;8 47–54.
10. Subramanian V, Van't Dack L, Van Grieken R. Chemical composition of river sediments from the Indian sub-continent. Chemical Geology 1985;48 271–279.
11. Kumar A, Kaur I, Mathur RP. Water quality and metal enrichment in bed sediment of the river Kali and Hindon, India. Environment and Geochemistry Health 1998;20 53–60.
12. Singh VK, Singh KP, Mohan M. Status of heavy metals in water and bed sediments of river Gomti – a tributary of the Ganga river,

India. Environmental Monitoring and Assessment 2005;105 43–67.

13. Abida B, Ramaiah M, Khan I, Veena K. Heavy Metal Pollution and Chemical Profile of Cauvery River Water. E-Journal of Chemistry 2009;6(1) 47–52.
14. Pandey J, Shubhashish K, Pandey R. Heavy metal contamination of Ganga river at Varanasi in relation to atmospheric deposition. Tropical Ecology 2010;51(2) 365–373.
15. Daniel Pepper. India's rivers are drowning in pollution. Fortune (magazine); 2007.
16. Karthikeyan, A. "Failure of Ganga, Yamuna projects." (http://articles.timesofindia.indiatimes.com/2009-09-04/chennai/28093527_1_cooum-treatment-plants-sewage-treatment). (Assessed 9 september 2009)
17. "Inflow to Yamuna to be cleaned up at last" (http://www.expressindia.com/latest-news/inflow-to-yamuna-to-be-cleaned-upat-last/509240/). Indian Express. (Assessed 31 August 2009).
18. Crnković D, Ristić M, Antonović D. Distribution of HeavyMetals and Arsenic in Soils of Belgrade (Serbia and Montenegro). Soil and Sediment Contamination 2006;15 581–589.
19. Madrid L, Diaz-Barrientos E, Madrid F. Metals in urban soils of Sevilla: seasonal changes and relations with other soil components and plant contents. European Journal of Soil Sciences 2004;55 209–217.
20. Chirenje T, Ma LQ, Reeves M, Szulczewski M. Lead distribution in near surface soils of two Florida cities: Gainesville and Miami. Geoderma 2004;119 113–120.
21. Imperato M, Adamo P, Naimo D, Arienzo M, Stanzione D, Violante P. Spatial distribution of heavy metals in urban soils of Naples city (Italy). Environment Pollution 2003;124 247–256.
22. Tijhuis L, Brattli B, Saether OM. A geochemical survey of topsoil in the city of Oslo, Norway. Environmental Geochemistry and Health 2002;24 67–94.
23. Raskin I, Smith RD, Salt DE. Phytoremediation of metals: using plants to remove pollutants from the environment. Current Opinion in Biotechnology 1997;8 221–226.

24. Salt DE, Smith RD, Raskin I. Phytoremediation. Annual Review of Plant Physiology and Plant Molecular Biology 1998; 49 643–668.

25. Vangronsveld J, Cunningham SD. 1998. Introduction to the Concepts. In: Vangronsveld J, Cunningham SD. (ed.) Metal Contaminated Soils: In situ Inactivation and Phytorestoration. Georgetown, USA: RG Landes Co.; 1998. p1-15.

26. Blaylock MJ, Huang JW. 2000. Phytoextraction of metals. In: Raskin I, Ensley BD (eds.) Phytoremediation of toxic metals: using plants to clean-up the environment. New York: John Wiley & Sons Inc; 2000. p53–70.

27. Salt DE, Blaylock M, Kumar NPBA, Dushenkov V, Ensley D, Chet I, Raskin I. Phytoremediation: a novel strategy for the removal of toxic metals from the environment using plants. Biotechnology 1995;13 468–474.

28. D'Souza R, Varun M, Pratas J, Paul MS. Paul Spatial distribution of heavy metals in soil and flora associated with the glass industry in North Central India: Implications for Phytoremediation. Soil and Sediment Contamination 2013;22 1–20.

29. Varun M, D'Souza R, Pratas J, Paul MS. Metal contamination of soils and plants associated with the glass industry in North Central India: Prospects of phytoremediation. Environment Science and Pollution Research 2012;9(1) 269-281.

30. Pratas J, Favas PJC, D'Souza R, Varun M, Paul MS. Phytoremedial assessment of flora tolerant to heavy metals in the contaminated soils of an abandoned Pb mine in Central Portugal. Chemosphere 2013;90 2216–2225.

31. Yang B, Shu W, Ye Z, Lan C, Wong M. Growth and Metal Accumulation in Vetiver and two Sebania species on Lead/Zinc Mine Tailings. Chemosphere 2003;52 1593–1600.

32. Uraguchi S, Watanabe I, Yoshitomi A, Kiyono M, Kuno K. Characteristics of cadmium accumulation and tolerance in novel Cd-accumulating crops,*Avena strigosa* and *Crotalaria juncea*. Journal of Experimental Botany 2006; 57(12) 2955-2965.

33. Varun M, D'Souza R, Kumar D, Paul MS. Bioassay as monitoring system for lead phytoremediation through *Crinum asiaticum* L. Environment Monitoring and Assessment 2011a; 178(1-4) 373–381.

34. Varun M, D'Souza RJ, Pratas J, Paul MS. Evaluation of phytostabilization, a green technology to remove heavy metals from industrial sludge using*Typha latifolia* L. Biotechnology Bioinformatics and Bioengineering 2011b; 1(1) 137-145.
35. D'Souza R, Varun M, Masih J, Paul MS. Identification of *Calotropis procera* L. as a potential phytoaccumulator of heavy metals from contaminated soils in Urban North Central India. Journal of Hazardous Material 2010;184 457–464.
36. BIS. Drinking water specification. IS:10500:1991. New Delhi, India: Bureau of Indian Standard; 1991.
37. WHO (World Health Organisation). Guidelines for drinking water quality. Vol. 1, recommendations, seconded. World health organization, Geneva; 1993.
38. Parkpain P, Sreesai S, Delaune RD. Bioavailability of heavy metals in sewage sludge-amended Thai soils. Water, Air, and Soil Pollution 2000;122 163–182.
39. USEPA (United States Environmental Protection Agency). Current drinking water standard. http://www.epa.gov/safewater/mcl.html. (accessed 2002).
40. Azumi, D.S, Bichi, M.H. Industrial pollution and heavy metals profile of Challawa river in kano, Nigeria. Journal of Applied Sciences and Environmental Sanitation 2010;5(1) 23–29.
41. Rodríguez-Barroso MR, Benhamou M, El Moumni B, El Hatimi I, García-Morales JL. Evaluation of metal contamination in sediments from north of Morocco: geochemical and statistical approaches. Environmental Monitoring And Assessment 2009;159(1-4) 169–181.
42. Halamić J, Galović I, Šparica M. Heavy Metal (As, Cd, Cu, Hg, Pb and Zn) distribution in topsoil developed on alluvial sediments of the Drava and Sava Rivers in NW Croatia. Geologia Croatica 2003;56(2) 215–232.
43. Amore RR. Concentration, speciation, provenance and pollution levels of heavy metals in the bottom sediments of the pasig river. M.Sc. Thesis. University of the Philippines, Diliman, Quezon; 2000.
44. Middelkoop FI. Heavy metal pollution of the river Rhine and flood plains in The Netherlands, Netherlands. Journal of Geosciences 2000;79 411–428.

45. Chen HM, Zheng CR. Evaluation of purification ability of suspended substance with log ks for heavy metal in water bodies. Acta pedologica 1995;32 86–92.
46. Olivares-Rieumont S, de La Rosa D, Lima L, Graham DW, Alessandro KD, Borroto J, Martínez F, Sánchez J. Assessment of heavy metal levels in Almendares River sediments - Havana City, Cuba. Water Research 2005;39(16) 3945–3953.
47. Muniz P, Danulat E, Yanicelli B, García-Alonso J, Medina G, Bícego MC. Assessment of contamination by heavy metals and petroleum hydrocarbons in sediments of Montevideo harbor, Uruguay. Environment International 2004; 29 1019–1028.
48. Eysink GGJ, Padua HB, Bertoletti SAEP, Martins MC, Pereira DN, Roberto S. Heavy metals in the Vale do Ribeira and in Iguape-Cananéia CETESB. Ambiente 1988;2(1) 6–13. (in Portuguese)
49. Siqueira GW. Study of heavy metals levels and other elements in surface sediments at the Santos Estuarine System and Shelf Coastal of the Amazon. PhD thesis. São Paulo University, Instituto Oceanogr áfico, Brazil; 2003.
50. Milenkovic N, Damjanovic M, Ristic M. Study of Heavy Metal Pollution in Sediments from the Iron Gate (Danube River), Serbia and Montenegro. Polish Journal of Environmental Studies 2005;14(6) 781–787.
51. Akhabuhaya J, Lodenius M. Metal pollution of river msimbazi, Tanzania. Environment International 1988;14 511–514.
52. Mackey AP, Hodgkionson M, Nardella R. Nutrient levels and heavy metals in mangrove sediments from the Brisbane River, Australia. Marine Pollution Bulletin 1992;24(8) 418–420.
53. Charkhabi AH, Sakizadeh M, Bayat R. Land use effects on heavy metal pollution of river sediments in Guilan, southwest of the Caspian sea. Caspian Journal of Environmental Sciences 2008;6(2) 133–140.
54. Kucuksezgin F, Uluturhan E, Batki H. Distribution of heavy metals in water, particulate matter and sediments of Gediz River (Eastern Aegean). Environmental Monitoring and Assessment 2008;141 213–225.
55. Turekian K, Wedepohl KH. Distribution of the elements in some major units of the earth›s crust. Geological Society of American Bulletin 1961;72 175–192.

56. Meybeck M, Helmer R. The quality of rivers: From pristine stage to global pollution. Paleography, Paleoclimatology, and Paleoecology 1989;75 283–309.
57. Schiller AM, Boyle EA. Variability of dissolved trace metals in the Missisipi river. Geochimistry and Cosmochim. Acta 1987;51 3273–327.
58. CCME (Canadian environmental quality guidelines). The Canadian Council of Ministers of the Environment; 2003.
59. VROM (Dutch Ministry of Housing, Spatial Planning and the Environment) Ministerial circular on target and intervention values for soil remediation. Annex A: target values, soil remediation intervention values and indicative levels for serious contamination, DBO/1999226863, 2004; February 4.
60. SEPAC (State Environmental Protection Administration of China). Chinese Environmental Quality Standard for Soils; 1995; (GB15618-1995).
61. Skowronski GA, Seide M, Abdel-Rahman MS. Oral bioaccessibility of trivalent and hexavalent chromium in soil by simulated gastric fluid. Journal of Toxicology and Environmental Health-Part A 2001;63(5) 351–362.
62. Onianwa PC. Monitoring atmospheric metal pollution: a review of the use of mosses as indicators. Environmental Monitoring and Assessment 2001;71 13–5.
63. Zhang M, He Z, Calvert DV, Stoffella PJ, Jiang X. Surface runoff losses of copper and zinc in sandy soils. Journal of Environmental Quality 2003;32 909–915.
64. Adriano CD. Biogeochemistry of Trace Metals. United States: Lewis publishers; 1992.
65. Modis K, Komnitsas K. Optimum sampling density for the prediction of acid mine drainage in an underground sulphide mine. Mine Water Environment 2007;26 237–242.
66. Loska K, Wiechula D, Barska B, Cebula E, Chojnecka A. Assessment of arsenic enrichment of cultivated soils in Southern Poland. Polish Journal of Environmental Studies 2003;12(2) 187–192.
67. Maina HM, Barminas JT, Nkafamiya II. Levels and distribution of some heavy metals in soils in the vicinity of ashaka cement

factory, gombe state Nigeria. Journal of Chemical Society of Nigeria 2009;34 1–5.

68. Lone MI, He ZH, Peter J, Stoffella I, Yang X. Phytoremediation of heavy metal polluted soils and water: Progresses and perspectives. Journal of Zhejiang University Science B 2008;9(3) 210–220.
69. Almå AR, Kweyunga C, Manoko LK. Investigation of trace metal concentrations in soil, sediments and waters in the vicinity of "Geita Gold Mine" and "North Mara Gold Mine" in North West Tanzania (with Barrick Response). IPM-report; 2009.
70. Nijkamp P, Rodenburg AC, Wagtendonk JA. Success factors for sustainable urban brownfield development: A comparative study approach to polluted sites. Ecology Economics 2002;40 232–252.
71. Sánchez-Martin MJ, Garcia-Delgado M, Lorenzo LF, Rodriguez-Cruz MS, Arienzo M. Heavy metals in sewage sludge amended soils determined by sequential extractions as a function of incubation time of soils. Geoderma 2007; 142 262–273.
72. Manta DS, Angelone M, Bellanca A, Neri R, Sprovieri M. Heavy metals in urban soils: a case study from the city of Palermo (Sicily), Italy. Science of the Total Environment 2002;300 229–243.
73. Paterson E, Sanka M, Clark L. Urban soils as pollutant sinks-a case study from Aberdeen, Scotland. Applied Geochemistry 1996;11 129–131.
74. Chronopoulos J, Haidouti C, Chronopoulou-Sereli A, Massas I. Variations in plant and soil lead and cadmium content in urban parks in Athens. Science of the Total Environment 1997;196 91–98.
75. Thornton I. Metal contamination of soil in urban areas. In: Bullock P, Gregory PJ (ed) Soils in the Urban Environment. Blackwell. London: Oxford; 1991. p47–75.
76. De Miguel E, Jiménez de Grado M, Llamas JF, Martín-Dorado A, Mazadiego LF. The overlooked contribution of compost application to the trace element load in the urban soil of Madrid (Spain). Science of the Total Environment 1998;215 113–122.
77. Kahle P. Schwermetallstatus Rostocker Gartenböden. Journal of Plant Nutrition and Soil Science 2000;163 19–196.

78. Li XD, Lee SL, Wong SC, Shi WZ, Thornton I. The study of metal contamination in urban soils of Hong Kong using a GIS-based approach. Environment Pollution 2004;129(1) 113–124.
79. Lu Y, Gong ZT, Zhang GL, Burghardt W. Concentrations and chemical speciation of Cu, Zn, Pb and Cr of urban soils in Nanjing, China. Geoderma 2003;115 101–111.
80. Sekhar KC, Chary NS, Kamala CT, Vairamani M, Anjaneyulu Y, Balaram V, Sorlie JE. Risk communications: around the world environmental risk assessment studies of heavy metal contamination in the industrial area of Kattedan, India—A case study. Human and Ecological Risk Assessment 2006;12 408–422.
81. Prashanthi V, Rao J, Sreenivasa K, Raju A. Soil pollution due to a land disposal of industrial effluents. Journal of Industrial Pollution 2001;17 9–18.
82. Huang R. Environmental Pedology. Beijing: Higher Education Press; 1987.
83. Romkens PFAM, Salomons W. Cd, Cu, and Zn solubility in arable and forest soils: consequences of land use changes for metal mobility and risk assessment. Soil Science 1998;163(11) 859–871.
84. Viard B, Pihan F, Promeyrat S, Pihan JC. Integrated assessment of heavy metal (Pb, Zn, Cd) highway pollution: bioaccumulation in soil, Graminaceae and land snails. Chemosphere 2004;55 1349–1359.
85. Baker AJM, Brooks RR. Terrestrial higher plants which hyperaccumulate metallic elements- a review of their distribution, ecology and phytochemistry. Biorecovery 1989;1 81–126.
86. Baker AJM, McGrath SP, Reeves RD, Smith JAC. Metal hyperaccumulator plants: a review of the ecology and physiology of a biological resource for phytoremediation of metal-polluted soils. In: Terry N, Bañuelos G. (eds.) Phytoremediation of Contaminated Soil and Water. Boca Raton, FL, USA:CRC Press; 2000. p85–108.
87. Lasat MM. Phytoextraction of toxic metals: A review of biological mechanisms. Journal of Environmental Quality 2002; 31 109–120.

88. Saxena PK, Raj SK, Dan T, Perras MR, Vettakkorumakankav NN. Phytoremediation of metal contaminated and polluted soils. In: Prasad MNV, Hagemeyer J. (eds.) Heavy metal stress in plants – From molecules to ecosystems. Heidelberg, Berlin, New York: Springer–Verlag; 1999. p305–329.
89. Porębska G, Ostrowska A. Heavy metal accumulation in wild plants: implications for phytoremediation. Polish Journal of Environmental Studies 1999;8(6) 433–442.
90. Saraswat S, Rai JPN. Phytoextraction potential of six plant species grown in multimetal contaminated soil. Chemical ecology 2009;25 1–11.
91. Pulford ID, Watson C. Phytoremediation of heavy metal contaminated land by trees—a review. Environment International 2003;29(4) 529–540.
92. Chiu KK, Ye ZH, Wong MH. Growth of Vetiveria zizanioides and Phragmities australis on Pb/Zn and Cu mine tailings amended with manure compost and sewage sludge: a greenhouse study. Bioresource Technology 2006; 97 158–170.
93. Morgenthal TL, Cilliers SS, Kellner K, Hamburg HV, Michael MD. The vegetation of ash disposal sites at Hendrina Power Station II: Floristic composition. South African Journal of Botany 2001; 67(4) 520-532.
94. Oosting HJ. The study of Plant Communities. San Francisco: W.H. Freeman and Co; 1958.
95. Phillips EA. Methods of vegetation study. New York: Henry Holt & Co. Inc.; 1959.
96. Hanson HC, Churchill ED. The plant community. New York: Reinhold Publishing Corporation; 1961.
97. Curtis JT, McIntosh RP. An upland forest continuum in the prairie forest border region of Wisconsin. Ecology 1951; 32 476-496.

# Citations

## CHAPTER 1

Christine E Thuring and Nigel Dunnett, Vegetation Composition of Old Extensive Green Roofs (From 1980s Germany), doi:10.1186/2192-1709-3-4.

## CHAPTER 2

Reuben P Keller, Juergen Geist, Jonathan M Jeschke, and Ingolf Kühn, Invasive species in Europe: ecology, status, and policy, doi:10.1186/2190-4715-23-23.

# CHAPTER 3

Eberhard Weber, Envisioning South-South Relations in the Fields of Environmental Change and Migration in the Pacific Islands - Past, Present and Futures, doi:10.1186/s40728-014-0009-z.

# CHAPTER 4

Xiaohou Shao, Tingting Chang, Maomao Hou, Yalu Shao, and Jingnan Chen, Application of Active EM-Calcium in Green Agricultural Production — Case Study in Tomato and Flue-cured Tobacco Production, doi: 10.5772/58329.

# CHAPTER 5

Carlos Alberto Mathias Azania, Luciana Rossini Pinto, Rodrigo Cabral Adriano, Dilermando Perecin and Andréa Padua Azania (2013). The Use of Glyphosate in Sugarcane: A Brazilian Experience, Herbicides - Current Research and Case Studies in Use, Dr. Andrew Price (Ed.), ISBN: 978-953-51-1112-2, InTech, DOI: 10.5772/54958.

# CHAPTER 6

Manoj S. Paul, Mayank Varun, Rohan D'Souza, Paulo J.C. Favas and João Pratas (2014). Metal Contamination of Soils and Prospects of Phytoremediation in and Around River Yamuna: A Case Study from North-Central India, Environmental Risk Assessment of Soil Contamination, Dr. Maria C. Hernandez Soriano (Ed.), ISBN: 978-953-51-1235-8, InTech, DOI: 10.5772/57239.

# Index

**T**



**U**



**W**



**Y**